# ELEMENTARY CO-ORDINATE
# GEOMETRY

# ELEMENTARY CO-ORDINATE GEOMETRY

## GEOMETRY

*A book for beginners*

By

A. S. RAMSEY, M.A.

CAMBRIDGE
AT THE UNIVERSITY PRESS
1964

## CAMBRIDGE
### UNIVERSITY PRESS

University Printing House, Cambridge CB2 8BS, United Kingdom

Cambridge University Press is part of the University of Cambridge.

It furthers the University's mission by disseminating knowledge in the pursuit of education, learning and research at the highest international levels of excellence.

www.cambridge.org
Information on this title: www.cambridge.org/9781316612675

First edition 1932
Reprinted with corrections 1935, 1942, 1944, 1945
Second edition 1946
Reprinted 1948, 1950, 1956, 1960, 1964
First paperback edition 2016

A catalogue record for this publication is available from the British Library

ISBN 978-1-316-61267-5 Paperback

# CONTENTS

# PREFACE

This book is a companion volume to the author's *Elementary Calculus*. It is intended for beginners in general and in particular for a class of students who do not intend to become mathematical specialists but want to acquire a sound working knowledge of the elements of the subject.

It is hoped that the explanations will prove adequate and sufficiently simple. The book contains a full discussion of the subject up to conics referred to their axes, using both point equations and parametric methods wherever the latter are suitable; at the same time the text contains frequent sets of easy examples so that it can readily be adapted for shortened courses.

The straight line and circle are fully treated and there is an early chapter on loci, on tangents and normals to simple curves using the Calculus, and on the use of parametric methods. Then follow chapters on the conics including the rectangular hyperbola referred to its asymptotes and a short chapter on polar and pedal equations. Some harder examples are included in the sets at the ends of the chapters.

My thanks are gratefully tendered to my nephew, Mr F. A. Spencer, for reading the proofs and verifying the answers to examples.

A. S. R.

*Cambridge*
*August* 1932

# PREFACE TO THE SECOND EDITION

In order to provide still further for the needs of beginners a classified collection of easy Examples has been inserted at the end of the book. A few harder Examples have also been added at the ends of the chapters.

A. S. R.

*July* 1946.

# CHAPTER I

# INTRODUCTION

**1.1** Co-ordinate Geometry is a method of proving geometrical theorems by algebraical processes. If we limit our considerations to plane figures then geometry has to do with the positions of points and lines in a plane, and the position of a point is determined by its distances from two fixed lines in the plane. These distances can be measured numerically in terms of a given unit or can be denoted by algebraical symbols called the co-ordinates of the point. The geometrical relations between the positions of points then become expressible as algebraical relations between these co-ordinates. This will soon become clear as we proceed further.

**1.2 Co-ordinate axes.** Let $X'OX$ and $Y'OY$ be two lines at right angles, the former of which is drawn from left to right and the latter upwards on the paper. These are called the *co-ordinate axes* and $O$ is the *origin of co-ordinates*.

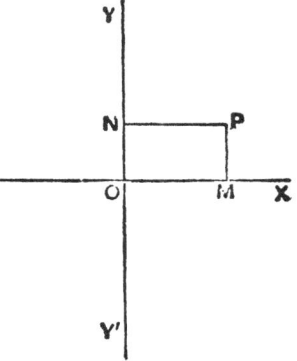

The position of any point $P$ is determined by its distances $NP$, $MP$ from the axes. These distances are called its *co-ordinates*. $NP$ parallel to $OX$ is denoted by $x$, and $MP$ parallel to $OY$ is denoted by $y^*$.

A definite point $P$ is reached from the origin by measuring along $OX$ a length $OM$ equal to a given value of $x$, and then measuring a length $MP$ parallel to $OY$ equal to a given value of $y$. $OM$ or $x$ is called the *abscissa* of $P$ and $MP$ or $y$ is called its *ordinate*.

---

* These co-ordinates are commonly called Cartesian after René Descartes (1596–1650) who made systematic use of them in his treatise on *Geometry* (1637).

The co-ordinates $x$ and $y$ can have any numerical values, positive or negative, negative values of $x$ being measured along $OX'$, and negative values of $y$ parallel to $OY'$.

It is usual to speak of the point $(x, y)$ meaning thereby the point whose co-ordinates are $x$ and $y$.

**1.21** In the accompanying figure the points marked *A*, *B*, *C*, *D* are the points $(4, 4)$, $(-4, 7)$, $(-8, -5)$, $(7, -7)$. As an exercise the student can write down the co-ordinates of the points *P*, *Q*, *R*, *S*.

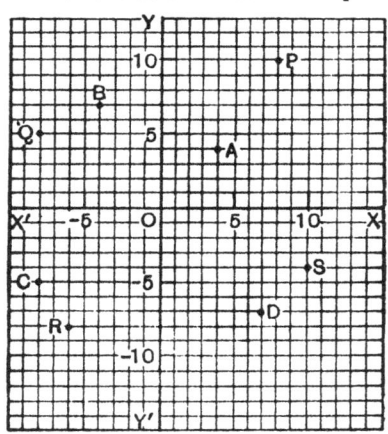

**1.3　Distance between two points.** The distance between a pair of points $P$, $Q$ of given co-ordinates $(x', y')$ and $(x'', y'')$ can be found by the theorem of Pythagoras.

Draw $PL$, $QM$, $QN$ parallel to the axes, as in the figure, so that

$$OL = x', \quad LP = y',$$
$$OM = x'', \quad MQ = y'',$$
then $\quad QN = ML = x' - x'',$
and $\quad NP = LP - MQ = y' - y''.$
Then $PQ^2 = QN^2 + NP^2$
$$= (x' - x'')^2 + (y' - y'')^2,$$
and $\qquad PQ = \sqrt{\{(x' - x'')^2 + (y' - y'')^2\}} \qquad \ldots\ldots(1).$

Cor. The distance of a point $P$, $(x', y')$, from the origin is given by $\qquad OP = \sqrt{(x'^2 + y'^2)}.$

**1.31 Example.** The distance between the points (9, 8) and $(-7, 4)$ is found directly from the formula 1.3 (1), by substituting $x' = 9$, $y' = 8$, $x'' = -7$, $y'' = 4$ and

$$= \sqrt{\{(9+7)^2 + (8-4)^2\}} = 4\sqrt{17}.$$

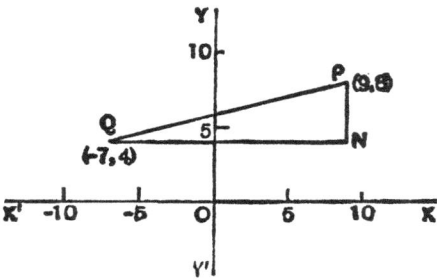

Or, without using the formula, we see from a figure that $QN = 16$ and $NP = 4$ and thus obtain $PQ$.

It is to be noted that the formulae of co-ordinate geometry are all *general* formulae applicable to points in all the quadrants provided that the proper signs are always affixed to the numerical values of the co-ordinates when they are introduced.

**1.32** In 1.3 the distance $PQ$ is found by extracting a square root, so that either sign might be given to it. Except in the case of lines parallel to either axis there is no convention about the directions to be considered positive or negative, and the distance between two points is measured as a positive number. But when we have three points $A$, $B$, $C$ in a straight line it is necessary to adhere to the rules

$$AB = -BA \quad \text{and} \quad AB + BC = AC,$$

no matter in what order the points are placed.

**1.4 Middle point of the line joining two given points.** Let $P$ be $(x', y')$ and $Q$ be $(x'', y'')$ and let the co-ordinates of $R$ the middle point of $PQ$ be $x$, $y$. Drawing parallels to the axes as in the figure we have, by parallels, that since $R$ is the

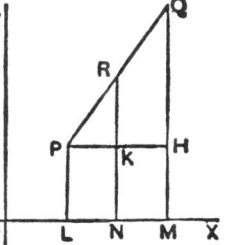

middle point of $PQ$, therefore $K$ is the middle point of $PH$ and $N$ is the middle point of $LM$ or
$$LN = NM,$$
but          $LN = x - x'$   and   $NM = x'' - x,$
therefore          $x - x' = x'' - x,$
and          $x = \frac{1}{2}(x' + x'').$

Similarly we can prove that
$$NR = \frac{1}{2}(LP + MQ),$$
or          $y = \frac{1}{2}(y' + y'').$

**1.5   Point dividing a line in a given ratio.** A like argument holds if the point $R$ is to divide $PQ$ in the ratio $m:n$, i.e. so that          $PR:RQ = m:n.$

For then by parallels
$$LN:NM = PK:KH = PR:RQ = m:n,$$
so that          $x - x' : x'' - x = m:n,$
or          $n(x - x') = m(x'' - x);$

and therefore          $x = \dfrac{nx' + mx''}{n + m},$

and similarly          $y = \dfrac{ny' + my''}{n + m}.$

We may use a single symbol to denote the ratio $m:n$, calling it the ratio $k:1$, say, then the co-ordinates of the point which divides the join of $(x', y')$ and $(x'', y'')$ in the ratio $k:1$ are
$$\frac{x' + kx''}{1 + k} \quad \text{and} \quad \frac{y' + ky''}{1 + k}.$$

**1.51** If we wish to divide a line *externally* in a given ratio the same formulae will hold good provided that the ratio $m:n$ or $k:1$ is a negative number.

Thus, if $R$ is a point on $PQ$ produced, we may say that $R$ divides $PQ$ externally in the ratio $PR:RQ$, and this is a negative number because $RQ$ has the opposite sign to $PR$. Also, if the numerical value of the ratio $PR:RQ$ is greater than 1, $R$ lies on $PQ$ produced, as above; but, if the numerical value is less than 1, then $R$ lies on $QP$ produced.

In numerical examples the word *externally* is to be taken as implying a negative sign for the ratio and when it is said that a line $PQ$ is divided internally and externally at $R$ and $R'$ in the ratio $m:n$ we must understand this to mean that

$$PR:RQ = m:n \quad \text{and} \quad PR':R'Q = -m:n.$$

The co-ordinates of $R$ are then given by 1.5, and the co-ordinates of $R'$ are given by

$$x = \frac{nx' - mx''}{n-m}, \quad y = \frac{ny' - my''}{n-m}.$$

**1.6 Area of a triangle.** Let $ABC$ be a triangle and let the co-ordinates of $A$, $B$, $C$ be $x_1, y_1; x_2, y_2; x_3, y_3$ respectively.

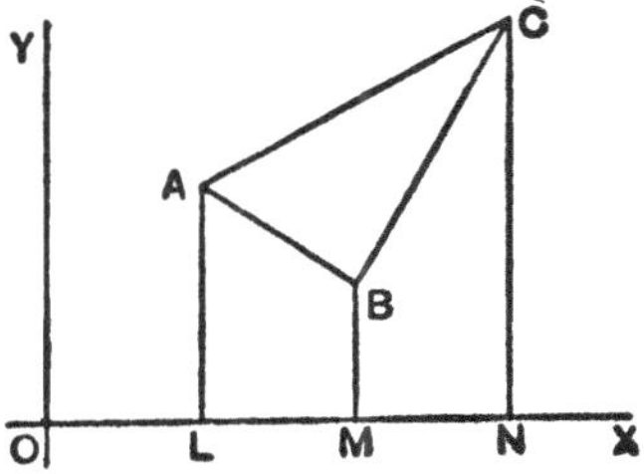

Draw $AL, BM, CN$ perpendicular to $OX$.

Then the area

$$ABC = ALNC - ALMB - BMNC.$$

But $ALNC = \triangle ALN + \triangle ACN$

$$= \tfrac{1}{2}LA \cdot LN + \tfrac{1}{2}LN \cdot NC$$

$$= \tfrac{1}{2}LN \,(LA + NC)$$

$$= \tfrac{1}{2}\,(x_3 - x_1)\,(y_1 + y_3).$$

Similarly $\qquad ALMB = \tfrac{1}{2}\,(x_2 - x_1)\,(y_1 + y_2),$

and $\qquad\qquad BMNC = \tfrac{1}{2}\,(x_3 - x_2)\,(y_2 + y_3).$

Therefore area

$$ABC = \tfrac{1}{2}\,\{(x_3 - x_1)\,(y_1 + y_3) - (x_2 - x_1)\,(y_1 + y_2) \\ - (x_3 - x_2)\,(y_2 + y_3)\},$$

which reduces to

$$\tfrac{1}{2}\,\{x_2 y_3 - x_3 y_2 + x_3 y_1 - x_1 y_3 + x_1 y_2 - x_2 y_1\}.$$

This will give a positive expression for the area if, as in the figure above, in going round the triangle $ABC$ in the order of the letters we have the area on the left hand; a negative answer would result from interchanging the order of two of the points.

The result can also be expressed as a determinant in the form

$$\tfrac{1}{2}\begin{vmatrix} x_1 & y_1 & 1 \\ x_2 & y_2 & 1 \\ x_3 & y_3 & 1 \end{vmatrix}.$$

**1.61 Examples.**

1. Mark on a diagram the points $(-2, 4)$, $(3, -5)$ and find the distance between them.

2. Find the lengths of the lines joining the following pairs of points:

$$(i)\ (-1, 1)\quad \text{and}\quad (2, -1);$$
$$(ii)\ (4, 3)\quad \text{and}\quad (-2, 2);$$
$$(iii)\ (a, b)\quad \text{and}\quad (-b, a).$$

3. Prove that the line joining the points $(2, 1)$ and $(4, 7)$ has the same middle point as the line joining $(5, 4)$ and $(1, 4)$, and hence that the four points are the corners of a parallelogram.

4. Prove that the points $(-3, -4)$, $(2, 6)$ and $(-6, 10)$ are at the corners of a right-angled triangle.

5. Plot the points $(0, 2)$, $(1, 1)$, $(4, 4)$ and $(3, 5)$ and prove that they are at the corners of a rectangle.

6. Plot the points $(-2, -2)$, $(-1, 2)$ and $(3, 1)$ and prove that they are at the corners of an isosceles triangle.

7. In the last question, find the distance of the vertex of the triangle from the middle point of the base.

8. Prove that the points $(-1, 0)$, $(0, 3)$, $(3, 2)$ and $(2, -1)$ are at the corners of a square.

9. Prove that the points $(-1, 0)$, $(3, 1)$, $(2, 2)$ and $(-2, 1)$ are at the corners of a parallelogram. By finding the co-ordinates verify that the joins of the middle points of pairs of opposite sides have the same middle point.

10. Prove that the points $(21, -2)$, $(15, 10)$, $(-5, 0)$ and $(1, -12)$ are at the corners of a rectangle, and find the co-ordinates of its centre.

11. Find the lengths of the medians of the triangle whose corners are at the points $(1, 2)$, $(0, 3)$ and $(-1, -2)$.

12. Find the co-ordinates of the points that divide the line joining the points $(-35, -20)$ and $(5, -10)$ into four equal parts.

13. Find the co-ordinates of the points of trisection of the line joining the points $(-5, -5)$ and $(25, 10)$.

14. Prove that the middle point of the line joining the points $(-5, 12)$ and $(9, -2)$ is a point of trisection of the line joining the points $(-8, -5)$ and $(7, 10)$.

**15.** The points $(8, 5)$, $(-7, -5)$ and $(-5, 5)$ are three of the corners of a parallelogram. Find the co-ordinates of the remaining corner which is to be taken as opposite to $(-7, -5)$.

**16.** The point $(2, 6)$ is the intersection of the diagonals of a parallelogram two of whose corners are at the points $(7, 16)$ and $(10, 2)$. Find the co-ordinates of the remaining corners.

**17.** Find the area of the triangle whose corners are the points $(2, 3)$, $(-4, 7)$, $(5, -2)$.

**18.** Find the co-ordinates of points which divide the join of $(2, 3)$, $(-4, 5)$ externally in the ratio $2:8$, and also externally in the ratio $8:2$.

**19.** Prove that if $(x_1, y_1)$, $(x_2, y_2)$, $(x_3, y_3)$ are the corners of a triangle the co-ordinates of its centroid are

$$\tfrac{1}{3}(x_1 + x_2 + x_3), \quad \tfrac{1}{3}(y_1 + y_2 + y_3).$$

**1.7 Loci.** When a point moves in a plane in conformity with some geometrical law its path in the plane is called its *locus*. Thus if a point moves so as to keep at a fixed distance from a given point its locus is a circle; if it moves so that its distances from two fixed points are always equal to one another its locus is the perpendicular bisector of the join of the two points, and so on.

In co-ordinate geometry loci are represented by equations. The co-ordinates of the moving point are denoted by $x$ and $y$, now called the "current co-ordinates," and the geometrical property in conformity with which the point moves is expressed symbolically as an equation between $x$ and $y$, which is called the equation of the locus.

For example:

(i) *If a circle is to be described with the point* $(2, 3)$ *as centre and with radius* 6, we have to express the fact that the point $(x, y)$ is at a distance 6 from the point $(2, 3)$, and from 1.3 it follows that

$$(x - 2)^2 + (y - 3)^2 = 36,$$

or $\qquad\qquad x^2 + y^2 - 4x - 6y - 23 = 0.$

We call this the equation of the circle.

(ii) *A point is to move so that it is at a constant distance a from the y-axis.*

If nothing more is specified, the point may be on either side of the $y$-axis, and therefore all points for which $x = a$, or $x = - a$ satisfy the required condition, and the whole locus may be represented by the single equation         $x^2 = a^2$.

(iii) *A point moves so that its distances from the points* (1, 2) *and* $(-2, -1)$ *are equal. Find the equation of the locus.*

If $(x, y)$ denotes any position of the point, we have from 1.3

$$(x - 1)^2 + (y - 2)^2 = (x + 2)^2 + (y + 1)^2,$$
which reduces to         $x + y = 0.$

**1.8**   We have seen in the foregoing examples that a single geometrical property about the motion of a point leads to an equation which represents the locus of the point. The method of co-ordinate geometry is to use some known fact about a curve in order to obtain its equation and then to deduce other properties of the curve from the equation so obtained.

**1.9   Examples.**

1.  A point moves so that its distance from the point (2, 1) is double its distance from the point (1, 2). Find the equation of its locus.

2.  Find the equation of the perpendicular bisector of the line joining the points $(3, -4)$ and $(-2, 3)$.

3.  Find the equation of the circle of radius 5 with centre at $(3, -4)$.

4.  A point moves so that its distance from the $y$-axis is equal to its distance from the point (2, 1). Find the equation of its locus.

5.  A point moves so that the sum of the squares of its distances from the points (3, 4) and (4, 3) is constant. Find the equation of the locus.

6.  A point moves so that its distance from the axis of $x$ is twice its distance from the point (0, 1). Find the equation of the locus.

7.  A point moves in such a way that with the points (2, 3) and $(-3, 4)$ it forms a triangle of area 8·5. Show that its locus has an equation         $(x + 5y)(x + 5y - 34) = 0.$

# CHAPTER II

# THE STRAIGHT LINE

**2.1 Straight lines parallel to an axis.** If a straight line is
parallel to the axis $OY$ every point on it
has the same abscissa, viz. $OM$, where $M$ is
the point in which the line cuts $OX$. Therefore if $OM = a$ the equation of the line is
$x = a$.

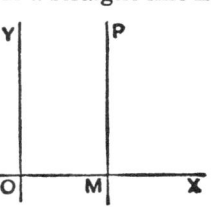

Conversely any equation of the form $x = a$
represents a straight line parallel to $OY$,
because every point on the locus is at the same distance from
$OY$.

Similarly a straight line parallel to $OX$ is represented by an
equation $y = b$.

**2.11 Straight lines through the origin.** Let $P$ be any
point on a straight line through $O$ which
makes an angle $\alpha$ with $OX$. Draw $PM$
perpendicular to $OX$ and let $x$, $y$ be
the co-ordinates of $P$. Then

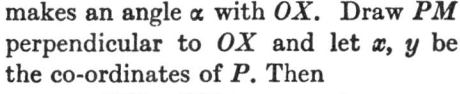

$$y = MP = OM \tan \alpha = x \tan \alpha.$$

Therefore $y = x \tan \alpha$
is the equation of the line.

We may also write $m$ for $\tan \alpha$ and then the equation is

$$y = mx,$$

and conversely this is the equation of a straight line through the
origin, $m$ denoting the tangent of the angle that the line makes
with $OX$.

**2.12 Equation of any straight
line.** Let $P$ be any point $(x, y)$ on
the straight line $TQP$ which cuts
$OY$ in $Q$ so that $OQ = c$ and makes
an angle $PTX = \alpha$ with $OX$.

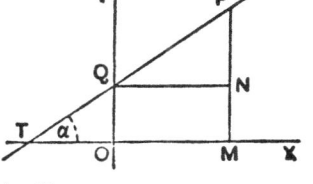

Draw $PM$ perpendicular to $OX$
and $QN$ parallel to $OX$ to meet $PM$ in $N$.

2

Then, by parallels, the angle

$$PQN = PTX = \alpha,$$

and
$$y = MP = NP + MN = QN \tan \alpha + MN$$
$$= OM \tan \alpha + OQ$$
$$= x \tan \alpha + c.$$

Therefore
$$y = x \tan \alpha + c$$

is the equation of the straight line or

$$y = mx + c \qquad\qquad \ldots\ldots(1),$$

if $m$ denotes the tangent of the angle that the line makes with $OX$.

2.121 It must be noted that the angle $\alpha$ is measured from the positive direction of $OX$ and that $c$ is positive when $OQ$ is positive.

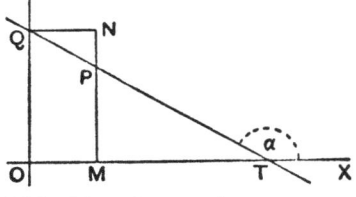

The adjoining figure shows a case in which $\alpha$ is obtuse. Here we have

$$y = MP = MN - PN$$
$$= MN - QN \tan(\pi - \alpha)$$
$$= c + OM \tan \alpha = c + x \tan \alpha,$$

or $y = mx + c$, as before.

The next figure shows a case in which $OQ = c$ is negative. Here we have

$$y = MP = NP - NM$$
$$= QN \tan \alpha + OQ$$
$$= x \tan \alpha + c,$$

or $y = mx + c$, as before.

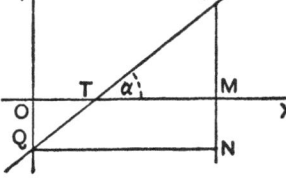

Cor. If the equations of two straight lines can be expressed in the forms

$$y = mx + c, \quad y = mx + c',$$

then the lines are parallel.

2.122 Examples.

(i) The equation of a straight line which makes an angle of 60° with $OX$ and cuts $OY$ at a distance of 2 units from $O$ is

$$y = \sqrt{3}x + 2.$$

(ii) By drawing a figure to show the data it is easy to see that we can find the equation of a straight line when we know the point in which it cuts either axis and its inclination to either axis. Thus, to find the equation of a straight line which cuts $OX$ at a distance 3 from the origin and makes an angle of 60° with $OY$.

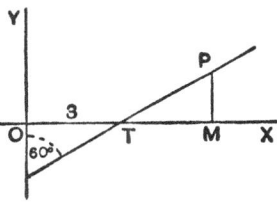

If $P$ be the point $(x, y)$ on the line which cuts $OX$ in $T$, and $MP$ be the ordinate of $P$, then $OT = 3$ and $OM = x$, therefore $TM = x - 3$, and the angle $PTM = 30°$; hence we have

$$y = MP = TM \tan PTM = (x - 3)/\sqrt{3}.$$

**2.123   Examples.** Find in each of the following cases the equation of the straight line which

(i) cuts $OY$ at distance 1 from $O$ and makes an angle of 45° with $OX$;

(ii)   ,,   ,,   $-3$   ,,   ,,   ,,   60°   ,,

(iii)   ,,   ,,   2   ,,   ,,   ,,   135°   ,,

(iv) cuts $OX$ at distance 2 from $O$ and makes an angle of 60° with $OX$;

(v)   ,,   ,,   $-2$   ,,   ,,   ,,   135°   ,,   $OY$;

(vi)   ,,   ,,   $-1$   ,,   ,,   ,,   135°   ,,   $OX$;

and draw figures showing the positions of the lines.

**2.2   Equation of a straight line in terms of the intercepts it makes on the axes.** Let $P$ be any point $(x, y)$ on the straight line $AB$ which cuts $OX$, $OY$ in $A$, $B$ so that $OA = a$ and $OB = b$.

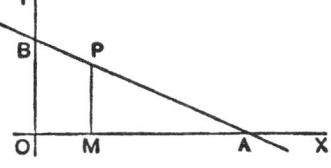

Then, if $MP$ is the ordinate of $P$, $OM = x$ and $MA = a - x$; but by similar triangles

$$\frac{MP}{MA} = \frac{OB}{OA},$$

therefore $\dfrac{y}{a - x} = \dfrac{b}{a}$, or $bx + ay = ab$.

By dividing by $ab$ we may write the equation

$$\frac{x}{a} + \frac{y}{b} = 1 \qquad \dots\dots(1).$$

This form admits of easy verification, for the locus of the equation cuts $OX$ where $y = 0$, i.e. where $x = a$, and cuts $OY$ where $x = 0$, i.e. where $y = b$.

The equation of a straight line may also be written

$$lx + my = 1 \qquad \ldots\ldots(2),$$

where $l$, $m$ are the reciprocals of the intercepts on the axes.

### 2.21 Examples.

1. Find the intercepts made on the axes by the straight lines whose equations are

(i) $2x + 8y = 12$;    (ii) $x - 1 = 8(y - 2)$;    (iii) $y = x$;

(iv) $\dfrac{x}{a+b} + \dfrac{y}{a-b} = \dfrac{1}{a^2 - b^2}$;    (v) $y = mx + c$.

2. Write down the equations of straight lines which make the following pairs of intercepts on the axes:

(i) 3, $-4$;    (ii) $-5$, 6;    (iii) $\dfrac{1}{a}$, $\dfrac{1}{b}$;    (iv) $2a$, $-2a$.

8. A straight line passes through a fixed point $(h, k)$ and cuts the axes in $A$, $B$. Parallels to the axes through $A$ and $B$ intersect in $P$. Find the equation of the locus of $P$.

### 2.3 An equation of the first degree represents a straight line.

In 2.12 and 2.2 we found that the equation of a straight line is of the first degree. We have now to prove the converse, that every equation of the first degree represents a straight line.

Let
$$Ax + By + C = 0 \qquad \ldots\ldots(1)$$

be the equation, and let $(x', y')$, $(x'', y'')$ be any two points $P$, $Q$ on the locus so that

$$\left. \begin{array}{l} Ax' + By' + C = 0 \\ Ax'' + By'' + C = 0 \end{array} \right\} \qquad \ldots\ldots(2).$$

and

A point $R$ which divides the join of $P$ and $Q$ in the ratio $k:1$ has co-ordinates $\dfrac{x' + kx''}{1+k}, \dfrac{y' + ky''}{1+k}$   (by **1.5**)

and this point also lies on the locus of the equation (1) if

$$\frac{A(x' + kx'')}{1+k} + \frac{B(y' + ky'')}{1+k} + C = 0,$$

or, clearing of fractions, if

$$Ax' + By' + C + k(Ax'' + By'' + C) = 0.$$

But because of (2) this condition is satisfied no matter what be the value of $k$. It follows that if any two points $P$, $Q$ are taken on the locus then every point $R$ on the straight line $PQ$ is also on the locus, therefore the locus is a straight line.

**2.4   Equation of a straight line in terms of the length of the perpendicular upon it from the origin and the angle which that perpendicular makes with an axis.** Let $OD = p$ be the perpendicular from the origin to the given line making an angle

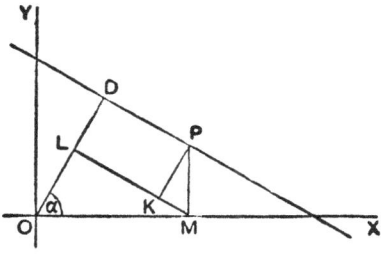

$$DOX = \alpha;$$

and let $P$ be any point $(x, y)$ on the line. Draw the ordinate $PM$, $ML$ parallel to $PD$ to meet $OD$ in $L$, and $PK$ perpendicular to $ML$.

Then
$$p = OD = OL + LD = OL + KP$$
$$= OM \cos \alpha + MP \sin \alpha,$$

or
$$p = x \cos \alpha + y \sin \alpha \qquad \qquad ......(1).$$

This is therefore the required equation.

**2.41**   The reader should draw figures with the line $OD$ falling in other quadrants and see that the equation holds in every case—for example in the fourth quadrant.

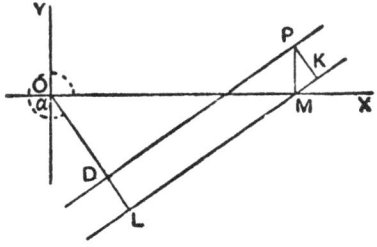

$$p = OD = OL - DL = OL - PK$$
$$= OM \cos (2\pi - \alpha) - MP \sin (2\pi - \alpha)$$
$$= x \cos \alpha + y \sin \alpha.$$

**2.42**   We have now obtained several forms for the equation of a straight line, viz.:

$$\text{(i) } y = mx + c \quad (2.12),$$

$$\text{(ii) } \frac{x}{a} + \frac{y}{b} = 1 \quad \text{and} \quad lx + my = 1 \quad (2.2),$$

$$\text{(iii) } Ax + By + C = 0 \quad (2.3),$$

and $\qquad$ (iv) $x \cos \alpha + y \sin \alpha = p \quad (2.4).$

It is a geometrical fact that a straight line can be drawn to satisfy *two* conditions, such as to pass through two assigned points, or to pass through a given point and make a given angle with a given direction; for this reason we expect the equation of a straight line to contain in general two independent constants, and this is so in the forms (i), (ii) and (iv) above, and also in (iii) because we can divide equation (iii) by any one of the constants and thus reduce their number to two. For example, dividing by $B$, we have $y = -\dfrac{A}{B} x - \dfrac{C}{B}$, which is the same as $y = mx + c$, if we put $m$ for $-A/B$ and $c$ for $-C/B$; or, dividing by $-C$, we have $-\dfrac{Ax}{C} - \dfrac{By}{C} = 1$, which is the same as $\dfrac{x}{a} + \dfrac{y}{b} = 1$, if we put $a$ for $-C/A$ and $b$ for $-C/B$.

It follows that the above forms are all equivalent; then (iii) can be expressed in the form (iv) by dividing by $\sqrt{(A^2 + B^2)}$ and writing

$$\frac{A}{\sqrt{(A^2 + B^2)}} x + \frac{B}{\sqrt{(A^2 + B^2)}} y = - \frac{C}{\sqrt{(A^2 + B^2)}},$$

and if we write $\cos \alpha$ for $A/\sqrt{(A^2 + B^2)}$ then $B/\sqrt{(A^2 + B^2)}$ is $\sin \alpha$, and denoting $-C/\sqrt{(A^2 + B^2)}$ by $p$, we have $x \cos \alpha + y \sin \alpha = p$.

Also by reference to 2.121 Cor. we see that two straight lines whose equations are

$$Ax + By + C = 0 \quad \text{and} \quad Ax + By + C' = 0,$$

are parallel, because the $m$ for both of them is $-A/B$.

### 2.43   Other forms of equation of a straight line.

(i) Let a line pass through a given point $A$ of co-ordinates $x'$, $y'$ and make an angle $\theta$ with $OX$.

Let $P$ be a point $(x, y)$ on the line at a distance $r$ from $A$. Draw $AM$, $PM$ parallel to $OX$, $YO$.

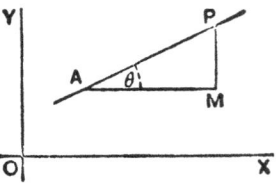

Then        $AM = AP \cos \theta$   and   $MP = AP \sin \theta,$

but          $AM = x - x'$   and   $MP = y - y',$

therefore    $x - x' = r \cos \theta$   and   $y - y' = r \sin \theta.$

Thus we have            $\dfrac{x - x'}{\cos \theta} = \dfrac{y - y'}{\sin \theta} = r$            ......(1)

as a symmetrical form of equations of a straight line.

If $\tan \theta = m$, we also have

$$y - y' = m\,(x - x')            ......(2),$$

but the form (1) is specially useful when we wish to bring into evidence the distance of the variable point $P$ from the fixed point $A$.

(ii) *To find the equation of a straight line which passes through two given points* $(x', y'), (x'', y'')$.

Let $A$, $B$ be the two given points $(x', y'), (x'', y'')$ and $P$ any point $(x, y)$ on the line. Draw $AM$, $BN$ parallel to $OX$, and $AN$, $PM$ parallel to $YO$, thus forming two similar triangles.

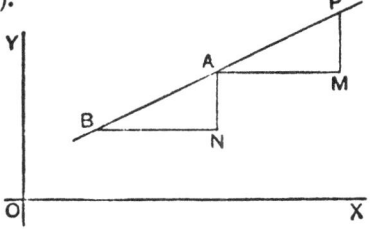

Then            $\dfrac{AM}{BN} = \dfrac{MP}{NA},$

but $AM = x - x',\ MP = y - y',\ BN = x' - x'',\ NA = y' - y'',$

therefore            $\dfrac{x - x'}{x' - x''} = \dfrac{y - y'}{y' - y''}$            ......(3)

is the equation required.

### 2.44 Examples.

(i) *Find the perpendicular distance of the line* $3x - 4y = 10$ *from the origin and the angle that the perpendicular makes with* $OX$.

Following the method of 2.42 we express the equation in the form $x \cos \alpha + y \sin \alpha = p$, by dividing it by $\sqrt{(3^2 + 4^2)}$ or 5, i.e. $\tfrac{3}{5}x - \tfrac{4}{5}y = 2$, and take $\cos \alpha = \tfrac{3}{5}$ and $\sin \alpha = -\tfrac{4}{5}$, then the required distance is given by $p = 2$. Also an angle whose cosine is positive and sine negative lies in the fourth quadrant and from tables we find

$$\alpha = 305° \ 51'.$$

It is easy to verify the position of the line and the perpendicular by noticing that the intercept on $OX$, found by putting $y = 0$, is $x = 3\frac{1}{3}$, and the intercept on $OY$, found by putting $x = 0$, is $-2\frac{1}{2}$.

(ii) *Find the equation of a straight line through the point* $(-2, 3)$ *making an angle of* $135°$ *with $OX$ and find its intercepts on the axes and its distance from the origin.*

Using 2.43 (2) the equation is

$$y - 3 = (x + 2) \tan 135°,$$

or 

$$y - 3 = -(x + 2),$$

or 

$$x + y = 1.$$

The intercepts on the axes are each 1, and by writing the equation

$$\frac{x}{\sqrt{2}} + \frac{y}{\sqrt{2}} = \frac{1}{\sqrt{2}}, \ \text{or} \ x \cos 45° + y \sin 45° = \frac{1}{\sqrt{2}},$$

it follows that the distance of the line from the origin is $1/\sqrt{2}$.

(iii) *Find the equation of a straight line through the point* $(-1, 5)$ *and having an intercept on $OX$ equal to twice its intercept on $OY$.*

Let $b$ be the intercept on $OY$, then $2b$ is the intercept on $OX$, and the equation of the line (2.2) is

$$\frac{x}{2b} + \frac{y}{b} = 1 \qquad \qquad ......(1),$$

but it passes through the point $(-1, 5)$ so that

$$-\frac{1}{2b} + \frac{5}{b} = 1,$$

whence $b = 9/2$, and by substituting in (1) we find that $x + 2y = 9$.

(iv) *Find the equation of a straight line through the point* $(2, 3)$ *and parallel to the line* $$5x - 7y + 4 = 0.$$

From the end of 2.42 we see that the equations of parallel straight lines only differ in their constant terms, so we may assume the required equation to be $$5x - 7y + c = 0,$$

since this is clearly a line parallel to

$$5x - 7y + 4 = 0.$$

Then by substituting for $x$, $y$ the co-ordinates 2, 3 of the point through which the line is to pass, we find that

$$10 - 21 + c = 0, \quad \text{or} \quad c = 11.$$

Therefore $\qquad 5x - 7y + 11 = 0$

is the required equation.

### 2.45 Examples.

1. Find the equations of two straight lines at a distance 3 from the origin and making an angle of 120° with $OX$.

2. Find the equation of a straight line making an angle of 60° with $OX$ and passing through the point $(2, -2)$.

Transform the equation to the form

$$x \cos \alpha + y \sin \alpha = p.$$

3. Find the equation of the straight line that passes through the points $(2, 3)$, $(3, 2)$. What is its inclination to $OX$?

4. Find the equation of the straight line through the point $(5, 7)$ that makes equal intercepts on the axes.

5. Find the equations of the sides of a triangle whose corners are $(2, 4)$, $(-4, 1)$, $(2, -3)$.

6. For the same triangle find the equations of the lines joining the corners to the middle points of the opposite sides.

7. Find the equation of a straight line passing through the point $(2, -3)$ parallel to the line $4x - y + 7 = 0$.

8. Find the intercepts on the axes made by a straight line which passes through the point $(3, -1)$ and makes an angle of 30° with $OX$.

9. Find the equation of the straight line through the points $(3, -4)$, $(2, 3)$, and of the parallel line through $(5, 2)$.

10. What is the distance from the origin of the line $4x - y = 7$? Write down the equation of a parallel line at double the distance.

11. Find the equation of the straight line through the point $(3, -4)$ parallel to the line joining the origin to the point $(2, -1)$.

12. Write down the equation of the straight line which makes intercepts 2 and $-7$ on the axes, and of the parallel line through the point $(3, -1)$.

13. Find the equations of the straight line joining the points $(3, 5)$, $(-2, 1)$ and of the parallel line through the origin.

**14.** *ABC* is a triangle and *A*, *B* and *C* are the points (2, 3), (5, − 1) and (− 4, 2). Find the equation of the straight line through *A* parallel to *BC*.

**15.** Find the equation of a line parallel to $2x + 5y = 11$ passing through the middle point of the join of the points (− 7, 8), (5, − 11).

**16.** The base of a triangle passes through a fixed point $(f, g)$ and the sides are bisected at right angles by the axes. Prove that the locus of the vertex is the line $gx + fy = 0$.

**2.5   Point of intersection of two straight lines.** Let the equations of the lines be

$$ax + by + c = 0,$$
$$a'x + b'y + c' = 0.$$

Their point of intersection has co-ordinates which satisfy both equations, so we have only to solve these simultaneous equations to find $x$ and $y$. The ordinary method of solution is to multiply the first equation by $b'$ and the second by $b$ and subtract, thus getting

$$(ab' - a'b)\,x + cb' - c'b = 0.$$

If in like manner we eliminate $x$ we get

$$(ab' - a'b)\,y + ac' - a'c = 0.$$

These two results may be written in the symmetrical form

$$\frac{x}{bc' - b'c} = \frac{y}{ca' - c'a} = \frac{1}{ab' - a'b}.$$

It should be noted that the denominators consist of the differences of cross products of the coefficients in the equations formed as indicated below

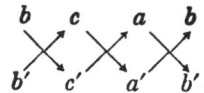

**2.51   Condition that three straight lines may meet in a point.** Let the equations of the lines be

$$ax + by + c = 0,$$
$$a'x + b'y + c' = 0,$$
$$a''x + b''y + c'' = 0.$$

The lines meet in a point if the point of intersection of two of them lies on the third. By **2.5** the first two meet in the point given by

$$\frac{x}{bc' - b'c} = \frac{y}{ca' - c'a} = \frac{1}{ab' - a'b},$$

and this point lies on the third if

$$\frac{a''\,(bc' - b'c)}{ab' - a'b} + \frac{b''\,(ca' - c'a)}{ab' - a'b} + c'' = 0,$$

or    $a''\,(bc' - b'c) + b''\,(ca' - c'a) + c''\,(ab' - a'b) = 0.$

This may be written as a determinant

$$\begin{vmatrix} a & b & c \\ a' & b' & c' \\ a'' & b'' & c'' \end{vmatrix} = 0.$$

The proof assumes that no two of the lines are parallel. If for example the first two lines were parallel, we should have $ab' - a'b = 0$ and their point of intersection would be at an infinite distance and the third line would only pass through this point if it were parallel to the other two.

**2.52    Angle between two straight lines of given equations.**

(i) If the equations of the lines are $y = mx + c$,   $y = m'x + c'$, then, if $\theta$, $\theta'$ are the angles that the lines make with $OX$, we have

$m = \tan \theta$ and $m' = \tan \theta'.$

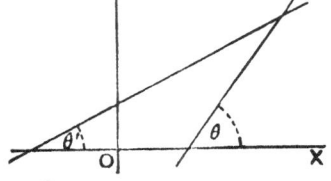

Hence    $\tan (\theta - \theta') = \dfrac{\tan \theta - \tan \theta'}{1 + \tan \theta \tan \theta'} = \dfrac{m - m'}{1 + mm'}.$

This gives one of the angles between the lines in the form $\tan^{-1} \dfrac{m - m'}{1 + mm'}$, and the other angle is its supplement. The sign of $(m - m')/(1 + mm')$ determines whether it represents the acute or the obtuse angle between the lines.

If $m = m'$ the lines are parallel.

If $1 + mm' = 0$ the lines are at right angles.

(ii) If the equations of the lines are

$$ax + by + c = 0 \quad \text{and} \quad a'x + b'y + c' = 0,$$

these can be written

$$y = -\frac{a}{b}x - \frac{c}{b} \text{ and } y = -\frac{a'}{b'}x - \frac{c'}{b'},$$

so that $\qquad m = -a/b \text{ and } m' = -a'/b',$

and the required angle is

$$\tan^{-1}\frac{-\dfrac{a}{b} + \dfrac{a'}{b'}}{1 + \dfrac{aa'}{bb'}} \text{ or } \tan^{-1}\frac{a'b - ab'}{aa' + bb'}.$$

We note that if $\dfrac{a}{a'} = \dfrac{b}{b'}$ the lines are parallel, and if $aa' + bb' = 0$ the lines are at right angles.

(iii) If the equations of the lines are

$$x \cos \alpha + y \sin \alpha = p \quad \text{and} \quad x \cos a' + y \sin \alpha' = p',$$

the perpendiculars on the lines from the origin make angles $\alpha$, $\alpha'$ with $OX$, therefore $\alpha \sim \alpha'$ is the angle between these perpendiculars. Therefore one of the angles between the given lines must be $\alpha \sim \alpha'$ and the other angle is its supplement.

**2.521**  In the last article we found the condition that two lines may be at right angles in the alternative forms

$$1 + mm' = 0 \quad \text{and} \quad aa' + bb' = 0.$$

In order to write down the equation of a line at right angles to a given line, we may proceed in either of two ways.

If the given line is $\qquad y = mx + c,$

the required equation is of the form

$$y = -\frac{x}{m} + c',$$

because $\qquad 1 + m \cdot \left(-\frac{1}{m}\right) = 0.$

If the given line is $\qquad ax + by + c = 0,$

the required equation is $bx - ay + c' = 0$

because $\qquad ab - ba = 0.$

In both cases there remains an unknown constant $c'$ which enables us to make the line pass through any assigned point.

Thus if we require a line to pass through $(x', y')$ and be at right angles to $$ax + by + c = 0,$$

we may write down $$bx - ay + c' = 0$$

to represent any straight line that satisfies the condition of perpendicularity, and then substitute $x = x'$, $y = y'$ and therefore

$$c' = -bx' + ay',$$

as the condition of passing through the assigned point, making the required equation

$$b(x - x') - a(y - y') = 0.$$

But we might have written this equation down at once, because from **2.43** any straight line through $(x', y')$ has an equation of the form

$$A(x - x') + B(y - y') = 0,$$

and it is only necessary to adjust the values of $A$ and $B$ so as to satisfy the condition of perpendicularity, i.e. write $b$ instead of $A$ and $-a$ instead of $B$.

**2.522 Example.** The equation of a line through $(2, 3)$ at right angles to the line $$7x - 4y + 8 = 0,$$

is $$4(x - 2) + 7(y - 3) = 0,$$

or $$4x + 7y - 29 = 0.$$

**2.53 Examples.**

1. Find the co-ordinates of the angular points of the triangle whose sides are $$3x + 2y + 6 = 0, \quad 2x - 5y + 4 = 0, \quad x - 3y - 6 = 0.$$

2. Prove that the lines

$$x + y + 25 = 0, \quad 2x + 3y + 7 = 0 \quad \text{and} \quad 3x + 5y = 11$$

are concurrent, and find the co-ordinates of their common point.

3. Find the equation of a line parallel to the line $2x - y = 3$ and passing through the intersection of the lines

$$3x + y = 7, \quad 3y = 2x - 5.$$

4. Find the equation of the line joining the origin to the point of intersection of the lines

$$3x - 5y = 11, \quad 2x + 7y + 4 = 0.$$

**5.** Find the acute angle between the lines

$$y = x + 7 \quad \text{and} \quad (2 + \sqrt{3})\, x + y = 11.$$

**6.** Find the angle between the lines

$$y = 2x + 5 \quad \text{and} \quad 2x + 4y + 11 = 0.$$

**7.** Find the equation of a straight line through the point $(2, -4)$ at right angles to the line $5x + 7y + 12 = 0$, and find the point in which the lines intersect.

**8.** Find the equation of a straight line through the origin and at right angles to the line $\quad ax + by + c = 0$.

**9.** Find the equation of a straight line at right angles to the line

$$5x - 2y + 11 = 0$$

and passing through the intersection of the lines

$$x + 2y + 1 = 0, \quad y = x + 7.$$

**10.** The origin is a corner of a square and two of its sides have equations $\quad y + 2x = 0, \quad y + 2x = 3.$
Find the equations of the other two sides.

**11.** Write down the equations of the perpendiculars from the origin to the lines $\quad x + 5y = 13, \quad 5x + y = 13$
and find the equation of the line joining the feet of the perpendiculars.

**12.** Prove that the line $x + y = 11$ makes equal angles with the lines

$$x - (2 - \sqrt{3})\, y + 2 = 0, \quad (2 - \sqrt{3})\, x - y + 5 = 0.$$

**13.** $A$ is the point $(-4, 0)$ and $B$ is the point $(3, 0)$. Find the locus of a point $P$ such that the angles $APO$, $OPB$ are equal, where $O$ is the origin.

**2.6** **The two sides of a line.** Let $y = mx + c$ be the equation of a straight line; and let the ordinate $QM$ of any point $Q\,(x', y')$ meet the line in $P$ whose co-ordinates are $(x', y'')$. Then since $P$ lies on the line therefore $\quad y'' = mx' + c.$

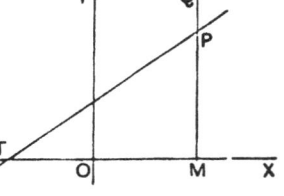

But $Q$ will be above or below the line according as $y'$ is $>$ or $< y''$, i.e. according as $y' >$ or $< mx' + c$ or according as $y' - mx' - c$ is positive or negative.

If we use any other form of equation for the straight line we have a like result and we conclude that if $ax + by + c = 0$ be the equation of a straight line, and the co-ordinates of a point $(x', y')$ *not* on the line be substituted in the expression

$$ax + by + c,$$

the result $\qquad ax' + by' + c$

will be positive for points lying on one side of the line and negative for points lying on the other side; and the two sides might be called the positive and negative sides of the line respectively. We note further that the origin lies on the positive or negative side of the line according as $c$ is positive or negative. If, however, we write the equation $-ax - by - c = 0$, then what we have previously called the positive and negative sides of the line now become interchanged.

2.61    For example consider the line $2x - 5y + 7 = 0$ and the points $O$ $(0, 0)$, $A$ $(1, 1)$, $B$ $(2, -1)$, $C$ $(-5, 1)$ and $D$ $(8, 4)$. The values of $2x - 5y + 7$ at the points $O, A, B, C, D$ are $7, 4, 16, -8, -7$; therefore $A$ and $B$ lie on the same side of the line as the origin and $C$ and $D$ lie on the opposite side.

### 2.62    The perpendicular distance of a point from a line.

(i) Let $\qquad x \cos \alpha + y \sin \alpha = p \qquad \ldots\ldots(1,$

be the equation of the given straight line $AB$; let $P$ be the given point having co-ordinates $x'$, $y'$, and let $d$ be the perpendicular distance $PM$ of $P$ from $AB$. Through $P$ draw a line $A'B'$ parallel

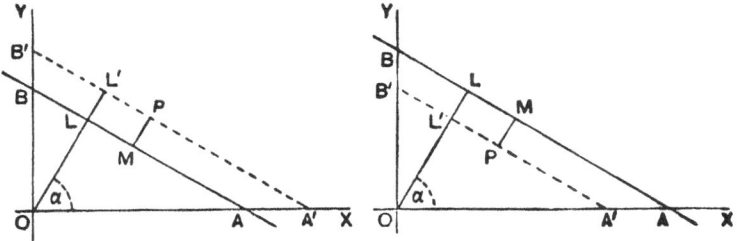

to $AB$ and let the perpendicular from $O$ to the parallels meet them in $L$, $L'$.

There are two cases; in the first $P$ is on the opposite side of $AB$ from the origin, and in the second $P$ and $O$ are on the same side of $AB$.

In both figures $OL = p$; in the first figure $OL' = p + d$ and in the second $OL' = p - d$. Therefore the equation of the line $A'B'$ is

$$x \cos \alpha + y \sin \alpha = p \pm d,$$

$\pm$ according as we look at the first or second figure. But $P$ lies on $A'B'$, therefore $x' \cos \alpha + y' \sin \alpha = p \pm d.$

Hence the required distance is given by

$$d = x' \cos \alpha + y' \sin \alpha - p,$$

when $P$ and $O$ are on opposite sides of the line; and by

$$d = p - x' \cos \alpha - y' \sin \alpha,$$

when $P$ and $O$ are on the same side of the line.

It follows that when the equation of a straight line is given in the form $x \cos \alpha + y \sin \alpha - p = 0,$

if the co-ordinates of any point not on the line are substituted for $x$ and $y$ in the expression on the left side of the equation the result represents numerically the distance of the point from the line. But it must be remembered that points on opposite sides of the line give opposite signs to the expression, and if we wish to represent the perpendicular distance of an assigned point as a positive length we must take note as to whether it lies on the same side of the line as the origin or not. If, as is usual, the perpendicular $p$ from the origin means a positive number, then $p - x \cos \alpha - y \sin \alpha$ will, as we have seen, be the form that will give positive lengths for perpendiculars from points on the same side of the line as the origin, and for points on the other side of the line the sign must be changed.

(ii) When the equation of the line is

$$ax + by + c = 0,$$

this can be changed to the form (i) by dividing it by $\sqrt{(a^2 + b^2)}$ (**2.42**); so that

$$\pm \frac{ax + by + c}{\sqrt{(a^2 + b^2)}} = 0$$

is the equation of the line in the form (1), and the perpendicular distance of the point $(x', y')$ is found by substituting $x'$, $y'$ for $x$, $y$ in the expression on the left side of the equation; i.e. the required distance is

$$\pm \frac{ax' + by' + c}{\sqrt{(a^2 + b^2)}}.$$

There is an ambiguity of sign here as in case (i) arising from the fact that $ax' + by' + c$, while keeping the same sign for all points on the same side of the line $ax + by + c = 0$, has opposite signs for points on opposite sides of the line.

The ambiguity is removed in any particular case in this way; the required perpendicular distance of the *assigned point* $P$ from the line is to be a *positive* number, so we first take note whether this point $P$ and the origin $O$ are on the same side or on opposite sides of the given line. If on the same side then the perpendicular from the origin is also to be a positive number, if on opposite sides then the perpendicular from the origin is to be a negative number. But, by putting $x' = y' = 0$, the above expression gives $\pm c/\sqrt{(a^2 + b^2)}$ as the length of the perpendicular from the origin, and we therefore choose the sign so as to make this a positive number or a negative number to suit the cases when $P$ and $O$ are on the same side or on opposite sides of the line respectively, and in each case this will secure that the perpendicular from $P$ is a positive number. A numerical example will make the process clear.

**2.621 Example.** *Find the centre and the radius of the inscribed circle of the triangle whose sides are the lines*

$$3x - 4y + 6 = 0, \ 5x + 12y - 72 = 0 \ and \ x + 5 = 0.$$

Let $AB$, $BC$, $CA$ be the three lines and $P$ the centre of the inscribed circle of the triangle $ABC$. Then we have to express the fact that $P$ is equidistant from the three lines.

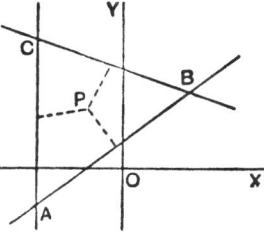

Let the co-ordinates of $P$ be $(x', y')$. The lengths of the perpendiculars from $P$ to the lines are

$$\pm \frac{(3x' - 4y' + 6)}{5}, \quad \pm \frac{(5x' + 12y' - 72)}{13}$$

and                $\pm (x' + 5),$

where the signs have to be determined so as to make all these lengths a positive number.

Now $P$ and $O$ are on opposite sides of $AB$ and $P$ is to be on the positive side, therefore $O$ is on the negative side and we must therefore take $- \dfrac{(3x' - 4y' + 6)}{5}$ for this will make the perpendicular from $O = - 6/5$.

Again $P$ and $O$ are on the same side of $BC$, so this must be chosen as the positive side for perpendiculars and we must take

$$- \frac{(5x' + 12y' - 72)}{13},$$

in order to get a positive number $72/13$ as the perpendicular from $O$. Lastly $P$ and $O$ are on the same side of $CA$, and therefore $+ (x' + 5)$ will make the perpendiculars from $P$ and $O$ both positive lengths.

Therefore $\quad -\dfrac{3x' - 4y' + 6}{5} = -\dfrac{5x' + 12y' - 72}{13} = x' + 5.$

Solving these equations we get $x' = -\frac{4}{2}\frac{3}{1}$, $y' = \frac{3}{8}\frac{0}{4}\frac{7}{4}$.

Then since each of the perpendicular distances that we have equated is a radius of the circle, we find by substituting for $x'$, $y'$ that the radius is $62/21$, and we also get a verification that we have arranged the signs correctly.

**2.63    To find the equations of the bisectors of the angles between two straight lines.** Let the equations of the given lines be $\qquad ax + by + c = 0 \quad$ and $\quad a'x + b'y + c' = 0.$

If a point lies on the bisector of either of the angles between the lines the perpendiculars from the point to the lines are of equal length. Therefore the point $(x, y)$ lies on a bisector if

$$\frac{ax + by + c}{\surd(a^2 + b^2)} = \pm \frac{a'x + b'y + c'}{\surd(a'^2 + b'^2)}.$$

These two equations, when we take the alternative signs, are the equations of the two bisectors. It remains to show how to distinguish between them.

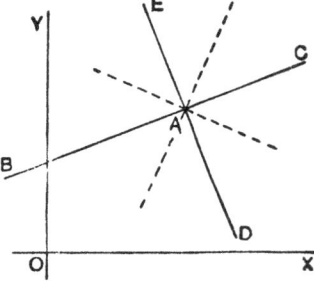

Let $BAC$ and $DAE$ be the two lines and let the origin fall within the angle $BAD$. Then the perpendiculars to the lines from any point within the angle $BAD$ will have the same signs as the perpendiculars from the origin.

Hence, if $c$, $c'$ are both positive numbers the expressions

$$ax + by + c \text{ and } a'x + b'y + c'$$

are both positive for points within the angle $BAD$ and the equation of the bisector of that angle is

$$\frac{ax + by + c}{\sqrt{(a^2 + b^2)}} = +\frac{a'x + b'y + c'}{\sqrt{(a'^2 + b'^2)}}.$$

Then, if we cross the line $AD$, $a'x + b'y + c'$ changes sign so that the equation of the bisector of the angle $DAC$ is

$$\frac{ax + by + c}{\sqrt{(a^2 + b^2)}} = -\frac{a'x + b'y + c'}{\sqrt{(a'^2 + b'^2)}}.$$

If $c$, $c'$ are both negative numbers a like argument gives the same equations for the bisectors of the two angles, but if $c$, $c'$ are one positive and the other negative then the signs in front of the right-hand side of the equations must be interchanged.

2.631 **Example.** *Find the bisectors of the angles between the lines* $3x + 4y + 7 = 0$, and $4x - 5 = 0$.

Let $BAC$ and $DAE$ be the lines. The equations of the bisectors are

$$\frac{3x + 4y + 7}{5} = \pm\frac{4x - 5}{4}.$$

In order that the perpendiculars from the origin to both the lines may be positive we must take the lower of the alternative signs. Then

$$\frac{3x + 4y + 7}{5} = -\frac{4x - 5}{4},$$

or $32x + 16y + 3 = 0$, is the bisector of the angle $CAE$ in the figure,

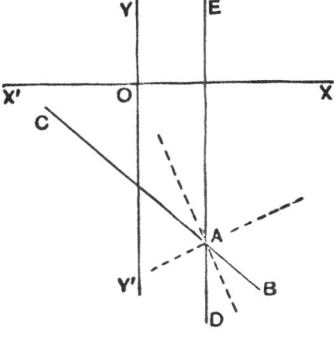

that being the angle that contains the origin; the equation

$$\frac{3x + 4y + 7}{5} = \frac{4x - 5}{4}, \text{ or } 8x - 16y - 53 = 0,$$

is the bisector of the angle $BAE$.

2.64 **Equation of a line through the point of intersection of two given lines.** Let $ax + by + c = 0$ ......(1)

and $a'x + b'y + c' = 0$ ......(2)

be the equations of the two given lines.

Consider the equation

$$ax + by + c + k\,(a'x + b'y + c') = 0 \qquad \ldots\ldots(3),$$

where $k$ is any constant.

It is an equation of the first degree and therefore represents a straight line. But if the co-ordinates of a point satisfy equations (1) and (2) they also satisfy (3), therefore (3) is a straight line through the common point of (1) and (2), and since it contains an arbitrary constant $k$ (i.e. a constant whose value may be chosen as may be convenient) it may be made to satisfy some one other condition.

Cor. Whenever the equation of a variable straight line can be expressed in the form (3) where $a$, $b$, $c$, $a'$, $b'$, $c'$ are definite constants and $k$ is a variable parameter then the line passes through a fixed point, namely the intersection of the fixed lines (1) and (2).

2.641    The last theorem is a special case of a more general theorem; that if $S = 0$ and $S' = 0$ are the equations of any two loci ($S$ and $S'$ denoting certain functions of $x$ and $y$), then the locus whose equation is $S + kS' = 0$ passes through all the points common to the loci $S = 0$ and $S' = 0$.

### 2.642    Examples.

(i) *Find the equation of a line through the point of intersection of the lines*
$$3x + y = 8, \quad 2x - y + 5 = 0,$$

*and parallel to the line*     $4x - 5y = 11.$

Any line through the point of intersection of the first two lines has equation     $3x + y - 8 + k\,(2x - y + 5) = 0,$

or     $(3 + 2k)\,x + (1 - k)\,y - 8 + 5k = 0 \qquad \ldots\ldots(1).$

This line will be parallel to $4x - 5y = 11$, if

$$\frac{3 + 2k}{4} = \frac{1 - k}{-5},$$

which gives $k = -\tfrac{19}{6}$; then substituting in (1) we get for the required equation     $20x - 25y + 143 = 0.$

By this method we avoid finding the co-ordinates of the point of intersection of the lines.

(ii) *Find the equation of the line joining the point* $(3, -5)$ *to the point of intersection of the lines*

$$2x - y + 7 = 0, \quad 5x + 2y - 11 = 0.$$

Any line through the point of intersection of the given lines has equation
$$2x - y + 7 + k(5x + 2y - 11) = 0,$$

and this passes through the point $(3, -5)$ if

$$6 + 5 + 7 + k(15 - 10 - 11) = 0, \quad \text{or} \quad k = 3.$$

Therefore the required line has equation

$$17x + 5y - 26 = 0.$$

**2.65**   *To prove that, if when the equations of three straight lines are multiplied by three constants and added the result is an identity, then the three straight lines have a common point.*

Let $ax + by + c = 0$, $a'x + b'y + c' = 0$ and $a''x + b''y + c'' = 0$ be the equations of the three lines, and suppose that constants $k, l, m$ can be found such that

$$k(ax + by + c) + l(a'x + b'y + c') + m(a''x + b''y + c'') = 0,$$

identically, i.e. no matter what values $x$ and $y$ may have, then the three lines are concurrent. This is evident if we consider that the point of intersection of the first two lines satisfies both

$$ax + by + c = 0 \quad \text{and} \quad a'x + b'y + c' = 0,$$

and if we substitute in the fourth equation it is therefore reduced to
$$m(a''x + b''y + c'') = 0,$$

so that the point of intersection of the first two lines also lies upon the third.

**2.651**   **Example.** *Prove that the perpendiculars from the corners of a triangle to the opposite sides meet in a point.*

Let $x', y'$; $x'', y''$; $x''', y'''$ be the co-ordinates of the corners $A, B, C$. The equation of $BC$ is

$$\frac{x - x''}{x'' - x'''} = \frac{y - y''}{y'' - y'''} \quad (2.43 \text{ (ii)}).$$

Then, by 2.521, the equation of a line through $A$ perpendicular to $BC$ is
$$(x'' - x''')(x - x') + (y'' - y''')(y - y') = 0.$$

Similarly the equations of the other two perpendiculars are

$$(x''' - x')(x - x'') + (y''' - y')(y - y'') = 0$$

and
$$(x' - x'')(x - x''') + (y' - y'')(y - y''') = 0.$$

Now if these equations are added together the sum of the left sides vanishes identically, therefore they represent three lines which meet in a point.

**2.66   Examples.**

**1.**  Find the distance of the point $(4, 2\sqrt{3})$ from the line

$$x \cos 60° + y \sin 60° = 6.$$

**2.**  Find the distance of the point $(2, 3)$ from each of the straight lines $5x + 12y - 20 = 0$, $4x - 3y + 11 = 0$, $3x + 4y - 28 = 0$.

**3.**  Find the distance of the point $(3, -1)$ from the line joining the points $(2, -3)$, $(4, 1)$.

**4.**  Are the points $(-2, 3)$, $(-2, 4)$ on the same or on opposite sides of the line $\qquad\qquad 4x + 5y = 10$?

**5.**  Find the equations of the bisectors of the angles between the lines $\qquad\qquad 2y = 4x - 9 \quad \text{and} \quad 2y = x + 4,$

and state which equation refers to the angle which contains the origin.

**6.**  Prove that the bisector of one of the angles between the lines

$$5x + y - 7 = 0 \quad \text{and} \quad x - 5y + 7 = 0$$

passes through the origin. What is the equation of the bisector of the other angle?

**7.**  What is the condition that the point $(x, y)$ may be at unit distance from the line $\qquad 3x - 4y + 10 = 0$?

Write down the equations of two straight lines parallel to the given line and at unit distances from it, and state which of the two lies on the same side of the given line as the origin.

**8.**  The sides $AB$, $BC$, $CA$ of a triangle have equations

$$4x - 3y = 12, \quad 3x + 4y = 24, \quad y = 2.$$

Find the co-ordinates of the centres of the inscribed circle and of the escribed circle opposite to the corner $A$.

**9.**  Prove that the point $(4, 4)$ lies outside the triangle whose sides are the lines

$$3x + 4y = 24, \quad 5x - 3y = 15, \quad y = 0.$$

**10.**  Find the equation of the line joining the origin to the point of intersection of the lines

$$x + 7y - 11 = 0, \quad y = 2x + 3.$$

11. Find the equation of a line perpendicular to the line

$$8x \quad 5y + 11 = 0$$

and passing through the intersection of the lines

$$5x - 6y = 1, \quad 3x + 2y + 5 = 0.$$

12. Find the equation of a line through the intersection of the lines

$$2x + 5y = 1 \quad \text{and} \quad y = 4x + 9$$

parallel to the line $\qquad x + y = 1.$

13. The corners of a triangle are at the points

$$(x', y'), \quad (x'', y''), \quad (x''', y''');$$

find the equations of the medians and prove that they meet in a point. What are the co-ordinates of their point of intersection?

14. For what multiples $k, l, m$ is the equation

$$k(2x + 8y - 13) + l(5x - y - 7) + m(x - 4y + 10) = 0$$

an identity?

In what point do the lines given by equating the three terms to zero concur?

15. Find the equations of the diagonals of the parallelogram

$$2x - y + 7 = 0, \quad 2x - y - 5 = 0, \quad 3x + 2y - 5 = 0, \quad 8x + 2y + 4 = 0.$$

16. The corners of a triangle are at the points

$$(2, 3), \quad (4, -3), \quad (-2, 1).$$

Find the equations of the perpendiculars to the sides through their middle points.

17. Work the same problem when the corners of the triangle are at the points $\qquad (x', y'), \quad (x'', y''), \quad (x''', y'''),$

and show that the perpendiculars meet in a point.

18. The line $\qquad 2x - 8y - 4 = 0$

is the perpendicular bisector of the line $AB$ and $A$ is the point $(5, 6)$; what are the co-ordinates of $B$?

## OTHER PROPERTIES OF THE EQUATIONS OF STRAIGHT LINES

**2.7   A homogeneous equation represents straight lines passing through the origin.** Let

$$a_0 y^n + a_1 x y^{n-1} + a_2 x^2 y^{n-2} + \ldots + a_n x^n = 0$$

be a homogeneous equation of the $n$th degree in $x$ and $y$ (i.e. the sum of the indices of $x$ and $y$ in every term is $n$). The left side of this equation can be resolved into $n$ real or imaginary factors

$$a_0 (y - m_1 x) (y - m_2 x) \ldots (y - m_n x) = 0,$$

and the equation is satisfied by the vanishing of any of its factors, and each of these equations

$$y - m_1 x = 0, \quad y - m_2 x = 0, \quad y - m_n x = 0$$

is a straight line through the origin.

For example, the equation

$$y^2 - 4xy + 3x^2 = 0$$

represents the two straight lines

$$y - x = 0 \quad \text{and} \quad y - 3x = 0.$$

The only real solution of the equation $x^2 + y^2 = 0$ is $x = 0$, $y = 0$, i.e. the origin. But the left side has imaginary factors, and we may regard the equation as representing two imaginary lines

$$x + \sqrt{(-1)}\, y = 0 \quad \text{and} \quad x - \sqrt{(-1)}\, y = 0$$

which have a real point of intersection $x = 0$, $y = 0$.

**2.71   To find the angle between the straight lines represented by the equation**

$$ax^2 + 2hxy + by^2 = 0.$$

Suppose that the equation is equivalent to

$$b (y - m_1 x) (y - m_2 x) = 0.$$

Then by multiplying out and comparing, we find that

$$- b (m_1 + m_2) = 2h \quad \text{and} \quad b m_1 m_2 = a.$$

Therefore   $(m_1 - m_2)^2 = \dfrac{4h^2}{b^2} - \dfrac{4a}{b} = \dfrac{4\,(h^2 - ab)}{b^2}.$

But the angle between the lines $y - m_1x = 0$ and $y - m_2x = 0$ is given by

$$\tan \theta = \frac{m_1 - m_2}{1 + m_1 m_2}$$

$$= \pm \frac{2\sqrt{(h^2 - ab)}}{a + b}.$$

We notice that the lines are only real and distinct when $h^2 > ab$, that they are real and coincident when $h^2 = ab$, and that they are at right angles when $a + b = 0$, i.e. when the sum of the coefficients of $x^2$ and $y^2$ is zero.

The last fact is a useful one to remember; e.g. the equation

$$2x^2 + 7xy - 2y^2 = 0$$

represents two straight lines at right angles.

### 2.72 Examples.

1. What lines are represented by the following equations:

   (i) $xy = 0$;  (ii) $x^2 - y^2 = 0$;

   (iii) $6x^2 + xy - y^2 = 0$;  (iv) $y^2 - 2xy \tan \theta - x^2 = 0$;

   (v) $x^3 + 3x^2y - 3xy^2 - y^3 = 0$?

2. Find the angles between the pairs of straight lines represented by the following equations:

   (i) $x^2 - 4xy + y^2 = 0$;  (ii) $x^2 - y^2 = 0$;

   (iii) $x^2 - 5xy + 4y^2 = 0$;  (iv) $x(x + 4y) = 0$;

   (v) $x^2 + xy - 6y^2 = 0$.

3. Prove that the equations

$$x^2 - 4xy + y^2 = 0 \quad \text{and} \quad x + y = 3$$

are the sides of an equilateral triangle.

### 2.8 The condition that the general equation of the second degree may represent two straight lines. 
It is clear that by multiplying together the equations of two straight lines, say

$$ax + by + c = 0 \quad \text{and} \quad a'x + b'y + c' = 0,$$

we get an equation of the second degree

$$(ax + by + c)(a'x + b'y + c') = 0,$$

which represents the two lines. But it does not follow that an equation

$$ax^2 + 2hxy + by^2 + 2gx + 2fy + c = 0 \qquad \ldots\ldots(1)$$

can be resolved into factors of the first degree, and unless this resolution is possible the equation (1) does not represent straight lines.

Suppose that $a$ is not zero. Then multiply the equation by $a$ and collect the terms in $x$ completing them into a square and setting the other terms on the other side of the equation. This gives

$$(ax + hy + g)^2 = (h^2 - ab)\,y^2 + 2\,(hg - af)\,y + g^2 - ac.$$

This equation can be resolved into linear equations if and only if the right-hand side is a perfect square, so that we can take the square roots of both sides. The condition for this is

$$(hg - af)^2 - (h^2 - ab)\,(g^2 - ac) = 0,$$

or, on multiplying out and dropping a factor $a$,

$$abc + 2fgh - af^2 - bg^2 - ch^2 = 0 \qquad \ldots\ldots(2).$$

If $a$ is zero and $b$ is not zero, we multiply by $b$ and complete the square in $y$ and obtain the same result without the terms in $a$.

If $a$ and $b$ are both zero, the equation is

$$2hxy + 2gx + 2fy + c = 0,$$

or $\qquad 2x\,(hy + g) + 2fy + c = 0;$

and for the left side to break into factors $hy + g$ must be a multiple of $2fy + c$, so that

$$\frac{h}{2f} = \frac{g}{c} \text{ or } ch = 2fg$$

is now the necessary and sufficient condition, being what the condition (2) reduces to when $a = b = 0$.

**2.81** It follows that in general the equation (1) of 2.8 does not represent straight lines but some other locus which we will call $S$. There is a simple method of finding the equations of the lines joining the origin to the intersections of $S$ by a straight line $lx + my = 1$. We use this equation and the equation of $S$ to make up a homogeneous equation, thus:

$$ax^2 + 2hxy + by^2 + 2\,(gx + fy)\,(lx + my) + c\,(lx + my)^2 = 0 \quad \ldots(3).$$

Now this equation being homogeneous represents straight lines passing through the origin, and to find where they meet the line $lx + my = 1$, we substitute 1 for $lx + my$ in (3) and the result is the equation of $S$, therefore (3) is the equation of the lines joining the origin to the points in which $S$ is cut by the line $lx + my = 1$.

**2.9 Miscellaneous examples.**

1. Find the locus of a point which is equidistant from the points $(6, -3)$, $(-4, 7)$.

2. Find the point on the line

$$2x + 5y + 7 = 0$$

which is equidistant from the points $(2, -3)$, $(-4, 1)$.

3. Find the co-ordinates of the circumcentre of the triangle whose corners are at the points $(4, 3)$, $(-1, 2)$, $(2, -2)$.

4. Find the equations of the lines through $(3, 1)$ which are respectively parallel and perpendicular to the line joining the points $(2, 4)$, $(5, -6)$.

5. Find the locus of a point at which the join of the points $(2, 1)$ and $(-3, 4)$ subtends a right angle.

6. Find the orthocentre of a triangle whose corners are at the points $(1, 2)$, $(-3, -4)$, $(6, 2)$.

7. Prove that the line joining the points $(2, -1)$, $(-3, 5)$ makes with the axes a triangle of area $\frac{49}{60}$.

8. $ABCD$ is a parallelogram and the co-ordinates of $A$, $B$ and $C$ are $2, 4; 1, 2$ and $4, 1$. Find the co-ordinates of $D$.

9. Find the area of the triangle formed by the lines

$$3x - 2y = 5, \quad 3y + 4x = 7, \quad y + 2 = 0.$$

10. Find the centre of the inscribed circle of the triangle whose sides are $\quad 3x - 4y = 0, \quad 12x - 5y = 0, \quad 4x + 3y = 8.$

11. The ends of a diagonal of a square are on the co-ordinate axes at the points $(2a, 0)$, $(0, a)$. Find the equations of the sides.

12. The sides of a triangle $ABC$ are

$$AB = 3, \quad BC = 5, \quad CA = 4,$$

and $A$, $B$ are on the axes $OX$, $OY$ respectively, while $AC$ makes an angle $\theta$ with $OX$. Prove that the locus of $C$, as $\theta$ varies, is given by the equation $\quad 16x^2 - 24xy + 25y^2 = 256.$

13. Prove that the locus of a point at which the join of the points $(a, 0)$ and $(-a, 0)$ subtends an angle of $45°$ is

$$x^2 + y^2 - 2ay = a^2.$$

**14.** Prove that the line $ax + by + c = 0$
divides the join of the points $(x', y')$, $(x'', y'')$ in the ratio
$$- (ax' + by' + c)/(ax'' + by'' + c).$$

**15.** Find the equation of the line joining the point $(x', y')$ to the point of intersection of the lines
$$ax + by + c = 0, \quad a'x + b'y + c' = 0.$$

**16.** Prove that the equations of the diagonals of the parallelogram whose sides are
$$ax + by + c = 0, \quad ax + by + d = 0,$$
$$a'x + b'y + c' = 0, \quad a'x + b'y + d' = 0,$$
are
$$(c' - d')(ax + by + c) - (c - d)(a'x + b'y + c') = 0$$
and
$$(c' - d')(ax + by + c) + (c - d)(a'x + b'y + d') = 0.$$
$$(c - d)(c' - d')/(ab' - a'b).$$

**18.** Prove that for all values of $k$ the line
$$(2 + k)x + (1 - 2k)y + 5 = 0$$
passes through a fixed point, and find its co-ordinates.

**19.** Find the angle between the lines
$$x^2 - 2xy \sec \theta + y^2 = 0.$$

**20.** Prove that the pairs of straight lines represented by
$$x^2 + xy = 0, \quad 6x^2 - xy - y^2 = 0$$
are such that the angles between one pair are equal to the angles between the other pair.

**21.** Find the angles between the lines
$$x^3 - 3x^2y - 3xy^2 + y^3 = 0.$$

**22.** Find the area of the triangle whose sides are given by
$$x^2 + 3y^2 = 4xy \quad \text{and} \quad 3x + 4y = 7.$$

**23.** Show that the equation
$$6x^2 - xy - 15y^2 - 11x + 31y - 10 = 0$$
represents two straight lines, and find the equations of the bisectors of the angles between them.

**24.** For what value of $k$ does the equation
$$12x^2 + 7xy + ky^2 + 13x - y + 3 = 0$$
represent two straight lines? What is the angle between them?

**25.** For what values of $k$ does the equation
$$6x^2 + kxy - 3y^2 + 4x + 5y - 2 = 0$$
represent two straight lines?

# CHAPTER III

# CURVES

**3.1** In this chapter we shall consider some examples of finding the equations of curves regarded as the loci of moving points, and then some properties of curves deduced from their equations.

### 3.11 Examples of loci.

(i) *A triangle has a given base and its sides are in a given ratio, find the locus of the vertex. (Circle of Apollonius.)*

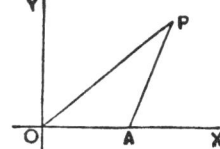

Let $OA = a$ be the base. Take $O$ for origin and $OX$ along $OA$. Then if $x$, $y$ are the co-ordinates of any position $P$ of the vertex such that $OP = k.AP$, we have

$OP^2 = k^2 AP^2$, or $x^2 + y^2 = k^2 \{(x - a)^2 + y^2\}$.

This may be written

$(1 - k^2)(x^2 + y^2) + 2ak^2x - k^2a^2 = 0$

and, as we shall see in the next chapter, represents a circle.

(ii) *Find the locus of a point such that its distance from a fixed point $(x', y')$ is proportional to its distance from a fixed line*

$$x \cos \alpha + y \sin \alpha - p = 0.$$

The required equation is

$$(x - x')^2 + (y - y')^2 = e^2 (x \cos \alpha + y \sin \alpha - p)^2,$$

where $e$ is the constant ratio of the distances.

(iii) *A point moves so that its co-ordinates at time $t$ are given by*

$$x = x_0 + ut, \quad y = y_0 + vt - \tfrac{1}{2}gt^2.$$

*Find the locus of the point.*

The first equation gives $t = (x - x_0)/u$,

and by substituting this value for $t$ in the second equation we get

$$u^2 (y - y_0) = vu (x - x_0) - \tfrac{1}{2}g (x - x_0)^2$$

as the equation of the locus.

### 3.12 Examples.

1. *$AB$ is a line of given length $a + b$. The ends $A$, $B$ are variable points on the axes $OX$, $OY$. $P$ is the point on $AB$ such that $AP = b$ and $PB = a$. Find the locus of $P$.*

**2.** Given the base of a triangle and the sum of the squares of the sides, find the locus of the vertex. [In this and the following examples take the middle point of the base for origin and $OX$ along the base.]

**3.** Given the base of a triangle and the product of the tangents of the angles at the base, find the locus of the vertex.

**4.** Given the base of a triangle and the difference of the angles at the base, find the locus of the vertex.

**5.** $ABC$ is a given triangle and the points $A$, $B$ are variable points on $OX$, $OY$. Prove that the locus of $C$ is

$$\frac{x^2}{a^2} - \frac{2xy \sin C}{ab} + \frac{y^2}{b^2} = \cos^2 C.$$

**6.** The angles of a triangle $ABC$ are given. The corner $A$ is at the fixed point $(f, 0)$ and $B$ moves along $OY$. Prove that the locus of $C$ is the line
$$x \cot A = y - f \cot C.$$

**3.2   Gradient of a curve.** Let the equation of a curve be $y = f(x)$, where $f(x)$ denotes some function of $x$.

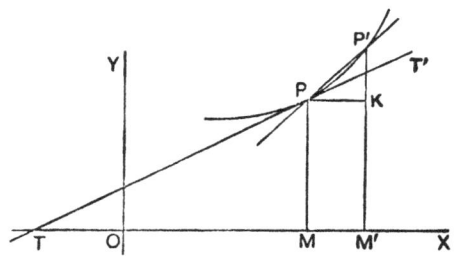

Let $P$ be the point $(x, y)$ on the curve and let $P'$ be a neighbouring point whose co-ordinates are $(x + \delta x, y + \delta y)$. $MP$, $M'P'$ are the ordinates at $P$, $P'$ and $PK$ parallel to $OX$ meets $M'P'$ in $K$.

Then $PK = \delta x$ and $KP' = \delta y$, and $\tan KPP' = \delta y/\delta x$.

Now the limiting position of the chord $PP'$ as $P'$ moves up to $P$ (i.e. as $\delta x \to 0$) is defined to be the tangent to the curve at $P$. Let $TPT'$ be this tangent. Then as $P'$ moves up to $P$, $\tan KPP'$ becomes $\tan MTP$, i.e. the tangent of the angle which the tangent to the curve at $P$ makes with $OX$.

Therefore $\quad \tan MTP = \underset{\delta x \to 0}{L} \dfrac{\delta y}{\delta x} = \dfrac{dy}{dx}$ or $f'(x)$.

This is called the *gradient of the curve at P*; i.e. the gradient of the curve at $P$ is the value of $dy/dx$ at the point $P$ on the curve*.

**3.21 Equation of the tangent to a curve** $y = f(x)$. Let it be required to find the equation of the tangent line to the curve $y = f(x)$ at the point $(x', y')$.

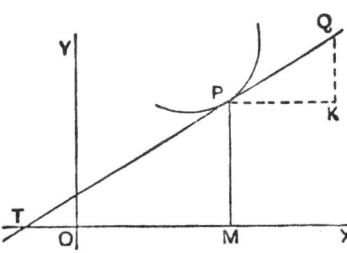

Let $P$ be the point $(x', y')$ and let $Q$ be any point $(x, y)$ on the tangent at $P$. By drawing lines through $P, Q$ parallel to the axes we make a right-angled triangle $PKQ$ whose sides are $PK = x - x'$, $KQ = y - y'$, so that

$$\frac{y - y'}{x - x'} = \tan QPK.$$

But $\tan QPK$ is the gradient of the curve at $P$ and therefore is equal to the value of $\dfrac{dy}{dx}$ at $P$, which we may denote either by $\left(\dfrac{dy}{dx}\right)_P$ or by $f'(x')$, the notation implying that we are to find $dy/dx$ from the equation of the curve and then put $x = x'$ to get the value of $dy/dx$ at $P$. Hence

$$y - y' = (x - x') \left(\frac{dy}{dx}\right)_P,$$

or $\qquad\qquad y - y' = (x - x') f'(x'),$

is the equation of the tangent to the curve at $(x', y')$.

**3.22 Examples.**

(i) *Find the equation of the tangent at the point $x = 2$ on the curve*

$$y = x^2 + 3x + 1.$$

Here $\dfrac{dy}{dx} = 2x + 3$; therefore, at $x = 2$, $y = 11$ and $\dfrac{dy}{dx} = 7$. Hence the equation of the tangent at $(2, 11)$ is

$$y - 11 = 7\,(x - 2), \text{ or } y - 7x + 3 = 0.$$

* See also *Calculus*, 3.1.

(ii) *Find the equations of the tangents to the curves*

$$y = x^2 + 2x \text{ and } y = x^2 - 5x + 7$$

*at the point where the curves intersect, and the angle at which they cut.*

The curves intersect where $x^2 + 2x = x^2 - 5x + 7$, i.e. in the point $x = 1, y = 3$.

On the first curve $\dfrac{dy}{dx} = 2x + 2, = 4$ at the point $x = 1$. Therefore the tangent at $(1, 3)$ to the first curve is

$$y - 3 = 4(x - 1), \text{ or } y = 4x - 1.$$

On the second curve $\dfrac{dy}{dx} = 2x - 5, = -3$ at the point $x = 1$. Therefore the tangent at $(1, 3)$ to the second curve is

$$y - 3 = -3(x - 1), \text{ or } y = -3x + 6.$$

The angle between these lines is

$$\tan^{-1} \frac{-3 - 4}{1 - 12} = \tan^{-1} \tfrac{7}{11} \quad (2.52).$$

(iii) *Find the equation of the tangent at the point $(x', y')$ on the curve* $xy = c^2$.

Since $y = \dfrac{c^2}{x}$, therefore $\dfrac{dy}{dx} = -\dfrac{c^2}{x^2}$, and at the point $(x', y')$ on the curve

$$\frac{dy}{dx} = -\frac{x'y'}{x'^2} = -\frac{y'}{x'}.$$

Therefore the equation of the tangent is

$$y - y' = -\frac{y'}{x'}(x - x'),$$

or $$x'y + y'x = 2x'y', \text{ or } x'y + y'x = 2c^2.$$

(iv) *Prove that the line $y = 3(x - 1)$ is a tangent to the curve*

$$y = x^4 - x.$$

The direct method of proof is to show that the line cuts the curve in two coincident points. At points common to the line and the curve we have

$$x^4 - x = 3x - 3, \text{ or } x^4 - 4x + 3 = 0 \qquad \ldots\ldots(1).$$

It is obvious that $x = 1$ satisfies the equation and by division that it is equivalent to

$$(x - 1)(x^3 + x^2 + x - 3) = 0.$$

Then we see that the factor $(x - 1)$ occurs again and that the equation is equivalent to

$$(x - 1)^2 (x^2 + 2x + 3) = 0.$$

Hence the line meets the curve in coincident points $x = 1$, $y = 0$ and is therefore a tangent.

If a root of (1) were not obvious we should apply the method of the Calculus to test for a repeated root[*].

### 3.23 Examples.

1. Find the equations of the tangents to the following curves at the points specified:

(i) $y = x(x^2 - 1)$ at $x = 2$;

(ii) $y = x^2 + \dfrac{1}{x^2}$ at $x = 1$;

(iii) $y = x^3 + 2x$ at $x = 0$;

(iv) $y = \left(x + \dfrac{1}{x}\right)^3$ at $x = 2$;

(v) $y = (x^2 - 1)^2$ at $x = 1$;

(vi) $y = x^3 - x + 1$ at $x = 3$;

(vii) $y = (x - a)^3$ at $x = 2a$;

(viii) $y = ax^2 + 2bx + c$ at $(x', y')$;

(ix) $y = \dfrac{x^3}{a^3} + \dfrac{a^3}{x^3}$ at $x = a$;

(x) $y = \dfrac{x^2}{a} + \dfrac{a^2}{x}$ at $x = a$.

2. Find the tangents to the curve $y = x^3 + x$ at the points where it is cut by the line $y = x + 4$ and find the point of intersection of the tangents.

3. Prove that the line $3x - 4y + 4 = 0$ touches the curve $y = x + \dfrac{1}{x}$.

4. Find the points on the curve $3y = x^3 + 3x$ at which the tangent is parallel to the line $y = 5x$.

5. Find at what points on the curve $y = x^2 + 9$ the tangents pass through the origin.

6. Show that there are three points on the curve

$$8y = 3x^4 + 8x^3 - 6x^2$$

at which the tangents are parallel to the line $y = 8x$.

7. Show that the line $4y = 17$ meets the curve $y = x^2 + \dfrac{1}{x^2}$ in four points and that two of the points of intersection of the tangents at these four points are on the line $4y + 1 = 0$, and two are on the line $x = 0$.

### 3.3 The tangent at a point on the curve $f(x, y) = 0$. An

equation $$f(x, y) = 0 \qquad \qquad \ldots\ldots(1)$$

between two variables may be regarded as defining one variable $y$ as a function of the other variable $x$, although it may not be

[*] *Calculus*, 3.422.

possible to solve the equation and express $y$ explicitly in terms of $x$. When such an equation represents a curve the gradient at any point $(x', y')$ is given by the value of $dy/dx$ at that point (3.2), where $dy/dx$ is to be found from $f(x, y) = 0$.

The result of differentiating this equation with regard to $x$ is

$$\frac{\partial f}{\partial x} + \frac{\partial f}{\partial y}\frac{dy}{dx} = 0 \qquad \qquad \dots\dots(2)^*,$$

where $\partial f/\partial x$ and $\partial f/\partial y$ are the partial derivatives of $f(x, y)$ with regard to $x$ and $y$.

Hence the gradient of the curve at $(x', y')$ is

$$\frac{dy}{dx} = -\frac{\partial f}{\partial x'}\bigg/\frac{\partial f}{\partial y'},$$

where $\dfrac{\partial f}{\partial x'}$ and $\dfrac{\partial f}{\partial y'}$ mean the values of $\dfrac{\partial f}{\partial x}$, $\dfrac{\partial f}{\partial y}$ at $(x', y')$.

Therefore the equation of the tangent to the curve $f(x, y) = 0$ at $(x', y')$ is

$$y - y' = -\frac{\dfrac{\partial f}{\partial x'}}{\dfrac{\partial f}{\partial y'}}(x - x'),$$

or

$$(x - x')\frac{\partial f}{\partial x'} + (y - y')\frac{\partial f}{\partial y'} = 0 \qquad \dots\dots(3).$$

**3.31    Example.** *Find the equation of the tangent at the point* $(x', y')$ *on the curve*

$$ax^2 + 2bxy + cy^2 - 1 = 0.$$

**Here**                $f(x, y) = ax^2 + 2bxy + cy^2 - 1,$

therefore        $\dfrac{\partial f}{\partial x} = 2ax + 2by$ and $\dfrac{\partial f}{\partial x'} = 2(ax' + by'),$

also            $\dfrac{\partial f}{\partial y} = 2bx + 2cy$ and $\dfrac{\partial f}{\partial y'} = 2(bx' + cy'),$

therefore the equation of the tangent at $(x', y')$ is

$$(x - x')(ax' + by') + (y - y')(bx' + cy') = 0,$$

or        $x(ax' + by') + y(bx' + cy') - ax'^2 - 2bx'y' - cy'^2 = 0,$

or            $x(ax' + by') + y(bx' + cy') - 1 = 0,$

since $(x', y')$ lies on the curve.

**3.32** The student who is unfamiliar with the subject of partial differentiation may omit for the present the formal method of the last two articles and use instead the following line of argument.

It is required to find the equation of the tangent at the point $(x', y')$ on the curve $$ax^2 + 2bxy + cy^2 - 1 = 0,$$
and for this purpose we want to find the value of $dy/dx$ at the point $(x', y')$. We regard the given equation as defining $y$ as a function of $x$ and differentiate the equation on this hypothesis. This means that if a term contains a power of $y$, say $y^n$, we must regard $y^n$ as a function of $y$ which is also a function of $x$ so that its derivative is

$$\frac{dy^n}{dy} \times \frac{dy}{dx} \text{ or } ny^{n-1}\frac{dy}{dx}.$$

The result of differentiating the above equation in this way is

$$2ax + 2by + 2bx\frac{dy}{dx} + 2cy\frac{dy}{dx} = 0,$$

therefore
$$\frac{dy}{dx} = -\frac{ax+by}{bx+cy}$$

and the value at $(x', y')$ is $\quad -\dfrac{ax'+by'}{bx'+cy'}.$

Hence the equation of the tangent at $(x', y')$ is

$$y - y' = -\frac{ax'+by'}{bx'+cy'}(x-x'),$$

or $\qquad (x - x')(ax' + by') + (y - y')(bx' + cy') = 0,$

or $\qquad x(ax' + by') + y(bx' + cy') - ax'^2 - 2bx'y' - cy'^2 = 0,$

or $\qquad x(ax' + by') + y(bx' + cy') - 1 = 0,$

since $(x', y')$ lies on the curve.

**3.321  Example.** *Find the equation of the tangent at the point $(x', y')$ on the curve $x^3 + y^3 - 3axy = 0$.*

Differentiating the equation gives

$$3x^2 + 3y^2\frac{dy}{dx} - 3ay - 3ax\frac{dy}{dx} = 0,$$

therefore $\dfrac{dy}{dx} = -\dfrac{x^2 - ay}{y^2 - ax}$, and the value at $(x', y')$ is $-\dfrac{x'^2 - ay'}{y'^2 - ax'}.$

Hence the equation of the tangent is

$$y - y' = -\frac{x'^2 - ay'}{y'^2 - ax'}(x - x'),$$

or $\qquad (x - x')(x'^2 - ay') + (y - y')(y'^2 - ax') = 0,$

or $\qquad x(x'^2 - ay') + y(y'^2 - ax') - x'^3 - y'^3 + 2ax'y' = 0.$

But $x'^2 + y'^2 = 3x'y'$, therefore the equation of the tangent is
$$x\,(x'^2 - ay') + y\,(y'^2 - ax') - ax'y' = 0.$$

### 3.33 Examples.

**1.** Find the equation of the tangents to the following curves at the points stated:

(i) $x^2 + y^2 = 25$, at $(3, 4)$;

(ii) $4x^2 + 9y^2 = 2$, at $(\frac{1}{2}, \frac{1}{3})$;

(iii) $y^2 = 4ax$, at $(a, 2a)$;

(iv) $\dfrac{x^2}{a^2} + \dfrac{y^2}{b^2} = 1$, at $(a\cos\theta, b\sin\theta)$;

(v) $x^2 - y^2 = a^2$, at $(a\sec\theta, a\tan\theta)$;

(vi) $x^2 - xy + y^2 = 4$, at $(2, 2)$;

(vii) $x^2 + y^2 + 2x + 4y - 20 = 0$, at $(3, 1)$;

(viii) $x^3 + y^3 - 3xy^2 + a^3 = 0$, at $(a, a)$.

**2.** Prove that the equation of the tangent at the point $(x', y')$ on each of the following curves has the form stated:

| Curve | Tangent |
|---|---|
| (i) $x^2 + y^2 = a^2$, | $xx' + yy' = a^2$; |
| (ii) $\dfrac{x^2}{a^2} + \dfrac{y^2}{b^2} = 1$, | $\dfrac{xx'}{a^2} + \dfrac{yy'}{b^2} = 1$; |
| (iii) $y^2 = 4ax$, | $yy' = 2a\,(x + x')$; |
| (iv) $xy = c^2$, | $xy' + yx' = 2c^2$; |
| (v) $y^2(x^2 - a^2) = a^2(x^2 + a^2)$, | $xx'(y'^2 - a^2) + yy'(x'^2 - a^2) = x'^2y'^2 + a^4$; |
| (vi) $x^{\frac{2}{3}} + y^{\frac{2}{3}} = a^{\frac{2}{3}}$, | $\dfrac{x}{x'^{\frac{1}{3}}} + \dfrac{y}{y'^{\frac{1}{3}}} = a^{\frac{2}{3}}$. |

### 3.4 Normals.

**DEF.** The **normal** at any point on a curve is the straight line through the point at right angles to the tangent at that point.

(i) *To find the equation of the normal at a point on the curve* $y = f(x)$.

Let $P$ be the point $(x', y')$ on the curve. Since the gradient or slope of the tangent at $P$ is $\left(\dfrac{dy}{dx}\right)_P$ or $f'(x')$ (3.2), therefore by

**2.521** the slope of the normal is $-1\Big/\left(\dfrac{dy}{dx}\right)_P$ or $-1/f'(x')$; **and the** normal is also a line through $(x', y')$, therefore its equation is

$$y - y' = - \frac{1}{\left(\dfrac{dy}{dx}\right)_P} (x - x'),$$

or $\qquad (x - x') + (y - y') \left(\dfrac{dy}{dx}\right)_P = 0 \qquad \ldots\ldots(1),$

or $\qquad (x - x') + (y - y') f'(x') = 0 \qquad \ldots\ldots(2).$

(ii) *To find the equation of the normal at a point on the curve* $f(x, y) = 0.$

We may assume the form given for the tangent in **3.3**, viz.

$$(x - x') \frac{\partial f}{\partial x'} + (y - y') \frac{\partial f}{\partial y'} = 0,$$

and then by **2.521** write down the equation of a line through $(x', y')$ at right angles to this, viz.

$$\frac{x - x'}{\dfrac{\partial f}{\partial x'}} = \frac{y - y'}{\dfrac{\partial f}{\partial y'}} \qquad \ldots\ldots(3),$$

which is then the required equation.

Or, alternatively, we may use the method of **3.32** to find from the equation $f(x, y) = 0$ the value of $dy/dx$ at $(x', y')$ and then substitute this value in equation (1) above.

**3.41 Examples.**

(i) *Find the equation of the normal at the point* (2, 6) *on the curve*
$$y = x^3 - x.$$

Here $\dfrac{dy}{dx} = 3x^2 - 1$, and the value at $x = 2$ is 11; therefore, by **3.4**(1) the normal at (2, 6) is

$$x - 2 + 11(y - 6) = 0 \text{ or } x + 11y - 68 = 0.$$

(ii) *Find the equation of the normal at the point* $(x', y')$ *on the curve*
$$x^3 + y^3 - 3xy^2 + a^3 = 0.$$

Here $\qquad f(x, y) = x^3 + y^3 - 3xy^2 + a^3,$

so that $\qquad \dfrac{\partial f}{\partial x} = 3(x^2 - y^2)$ and $\dfrac{\partial f}{\partial x'} = 3(x'^2 - y'^2);$

also $\qquad \dfrac{\partial f}{\partial y} = 3(y^2 - 2xy)$ and $\dfrac{\partial f}{\partial y'} = 3(y'^2 - 2x'y').$

Therefore, by 3.4 (3) the normal is

$$\frac{x-x'}{x'^2-y'^2} = \frac{y-y'}{y'^2-2x'y'}.$$

Otherwise, by differentiating the given equation, we get

$$3x^2 + 3y^2\frac{dy}{dx} - 3y^2 - 6xy\frac{dy}{dx} = 0,$$

therefore, at $(x', y')$, $\qquad \dfrac{dy}{dx} = -\dfrac{x'^2-y'^2}{y'^2-2x'y'},$

and then by 3.4 (1) the normal is

$$\frac{x-x'}{x'^2-y'^2} = \frac{y-y'}{y'^2-2x'y'}.$$

**3.42  Subtangent** and **subnormal.** If $NP$ be the ordinate
at a point $P$ on a curve and
the tangent and normal at $P$
meet the axis of $x$ in $T$ and $G$,
then $TN$ is called the **subtangent** and $NG$ is called the **subnormal at $P$.**

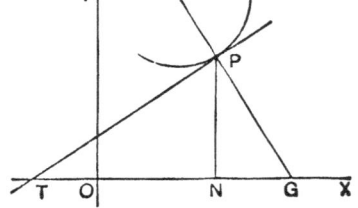

Since

$\tan NTP = dy/dx$ and $NP = y$,

therefore the subtangent $\qquad = y \Big/ \dfrac{dy}{dx},$

and the subnormal $\qquad = y\,\dfrac{dy}{dx}.$

**3.43  Examples.**

1. Find the equations of the normals to the following curves at
the points stated:

      (i) $x^2 + y^2 - 2x - 4y = 3$, at $(3, 4)$;

      (ii) $x^2 + 4xy + y^2 = 13$, at $(1, 2)$;

      (iii) $y^2 = 4ax$, at $(a, 2a)$;

      (iv) $\dfrac{x^2}{a^2} + \dfrac{y^2}{b^2} = 1$, at $(a\cos\theta, b\sin\theta)$;

      (v) $x^2 - 4y^2 = 4a^2$, at $(2a\sec\theta, a\tan\theta)$;

      (vi) $x^3 - y^3 - 3xy^2 + a^3 = 0$, at $(a, -a)$.

**2.** Prove that the equation of the normal at the point $(x', y')$ on each of the following curves has the form stated:

| Curve | Normal |
|-------|--------|
| (i) $x^2 + y^2 = a^2$, | $\dfrac{x}{x'} = \dfrac{y}{y'}$; |
| (ii) $\dfrac{x^2}{a^2} + \dfrac{y^2}{b^2} = 1$, | $\dfrac{x - x'}{\dfrac{x'}{a^2}} = \dfrac{y - y'}{\dfrac{y'}{b^2}}$; |
| (iii) $y^2 = 4ax$, | $\dfrac{x - x'}{2a} + \dfrac{y - y'}{y'} = 0$; |
| (iv) $xy = c^2$, | $\dfrac{x - x'}{y'} = \dfrac{y - y'}{x'}$. |

**3.** Prove that for the curve $y^2 = 4ax$ the subnormal is of constant length.

**4.** Prove that the portion of any tangent to the curve $x^{\frac{2}{3}} + y^{\frac{2}{3}} = a^{\frac{2}{3}}$ intercepted by the axes is of length $a$.

**5.** Prove that for the curve $ay^2 = x^3$ the subnormal varies as the square of the subtangent.

**6.** Prove that for the curve $y = ae^{\frac{x}{b}}$ the subtangent is of length $b$.

**7.** Prove that the area of the triangle formed by the axes and any tangent to the curve $xy = c^2$ is $2c^2$.

**8.** Prove that for the curve $x^m y^n = c^{m+n}$ the portion of a tangent intercepted by the axes is divided at the point of contact in the ratio $m : n$.

**9.** Prove that, if $N$ is the foot of the ordinate and $NT$ is the subtangent at a point on the curve $\dfrac{x^2}{a^2} + \dfrac{y^2}{b^2} = 1$, then $OT.ON = a^2$.

**10.** Prove that the perpendicular from the foot of the ordinate to the tangent to a curve is of length $y \bigg/ \sqrt{\left\{1 + \left(\dfrac{dy}{dx}\right)^2\right\}}$. Show that for the curve $y = c \cosh x/c$, this perpendicular is of length $c$.

**11.** Find the equation of the tangent to the curve
$$2x^3 + 2y^3 - 9axy = 0,$$
at the point $(2a, a)$; and show that the tangent meets the curve again where $\qquad 4x + y = 0.$

**3.5  Parametric representation of curves.** If a point is moving in a plane in such a way that its co-ordinates at any instant are both known functions of the time $t$, then it would be possible to find the equation of the path of the point by eliminating $t$ between the formulae that give $x$ and $y$ in terms of $t$.

For example, if the position of the point at time $t$ is given by

$$x = ut \quad \text{and} \quad y = vt - \tfrac{1}{2}gt^2 \qquad \ldots\ldots(1),$$

we deduce $y = \dfrac{v}{u}\,x - \dfrac{gx^2}{2u^2}$ as the equation of the path of the point.

Used in this sense $t$ is called a "variable parameter" and the curve is expressed parametrically by equations (1).

In a more general sense $t$ is not necessarily time but denotes a quantity having a field of variability which may include all numbers from $-\infty$ to $\infty$. The co-ordinates $x$, $y$ are said to be expressed parametrically by the relations

$$x = f(t), \quad y = g(t) \qquad \ldots\ldots(2),$$

where $f(t)$ and $g(t)$ are single-valued functions, so that to each value of $t$ there corresponds one value of $x$ and one of $y$, i.e. a single point on the curve.

**3.51  Examples.**

Find the equations in Cartesian co-ordinates of the curves given by the following pairs of parametric relations:

  (i) $x = a \cos t$;    $y = a \sin t$.      (ii) $x = at^2$;      $y = 2at$.

 (iii) $x = a \cos t$;    $y = b \sin t$.      (iv) $x = a \sec t$;    $y = b \tan t$.

  (v) $x = ct$;          $y = c/t$.        (vi) $x = a \cos^3 t$;    $y = b \sin^3 t$.

(vii) $x = a \cos t + b \cos 2t$;    $y = a \sin t + b \sin 2t$.

**3.52  Equation of the tangent and normal at a point on a curve given parametrically.** If the curve is given by equations $x = f(t)$, $y = g(t)$ then we may speak of a point $t'$ on the curve meaning thereby the point for which the parameter $t$ has the particular value $t'$; i.e. the point $x = f(t')$, $y = g(t')$.

To find the equation of the tangent at any point $t$, we have to find the gradient $dy/dx$ at this point. But in the notation of differentials $dx = f'(t)\,dt$ and $dy = g'(t)\,dt$, so that $\dfrac{dy}{dx} = \dfrac{g'(t)}{f'(t)}$.

Hence since the tangent also goes through the point whose co-ordinates are $f(t)$, $g(t)$ therefore its equation is

$$y - g(t) = \{(x - f(t)\} g'(t)/f'(t),$$

which may be written

$$\frac{x - f(t)}{f'(t)} = \frac{y - g(t)}{g'(t)}.$$

The normal being perpendicular to the tangent, its equation is

$$\{x - f(t)\} f'(t) + \{y - g(t)\} g'(t) = 0.$$

### 3.53 Examples.

1. *Find the equation of the tangent and normal at a point on the curve*

$$x = a + b \cos t, \quad y = b + a \sin t.$$

The equation of the tangent is

$$\frac{x - a - b \cos t}{-b \sin t} = \frac{y - b - a \sin t}{a \cos t},$$

or $\qquad (x - a) a \cos t + (y - b) b \sin t = ab;$

and that of the normal is

$$-b \sin t (x - a - b \cos t) + a \cos t (y - b - a \sin t) = 0,$$

or $\qquad (x - a) b \sin t - (y - b) a \cos t + (a^2 - b^2) \sin t \cos t = 0.$

2. Find in like manner the equations of tangents and normals to the curves whose parametric equations are given in 3.51.

### 3.6 Change of origin.

If, while keeping a figure fixed on the page we move the co-ordinate axes parallel to themselves through definite distances, the effect on the co-ordinates of every point in the figure is to increase or decrease every $x$ by the same amount and similarly every $y$.

Thus let $P$ be a point whose co-ordinates are $x$, $y$ referred to the axes $OX$, $OY$, and $x'$, $y'$ referred to the axes $O'X'$, $O'Y'$; where the co-ordinates of $O'$ referred to $OX$, $OY$ are $h$, $k$.

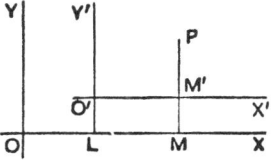

Then from the figure

$$x = OM = LM + OL = O'M' + OL = x' + h,$$

and $\qquad y = MP = M'P + MM' = M'P + LO' = y' + k.$

It follows that if the locus of $P$ referred to $OX$, $OY$ be given by the equation
$$f(x, y) = 0,$$
then referred to $O'X'$, $O'Y'$ the locus is
$$f(x' + h, y' + k) = 0;$$
i.e. the origin may be moved to the point $(h, k)$ while keeping the directions of the axes fixed if in every equation we increase every $x$ by $h$ and every $y$ by $k$.

**3.61  Rotation of axes.** Let the axes $OX$, $OY$ rotated through an angle $\theta$ in the sense from $OX$ towards $OY$ take the positions $OX'$, $OY'$.

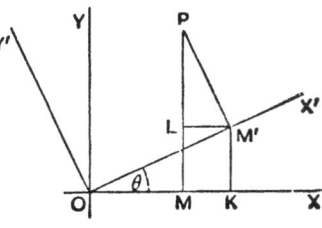

Let $x$, $y$ and $x'$, $y'$ be the co-ordinates of a point $P$ referred to the old and the new axes. Let $MP$, $M'P$ be the ordinates, and $M'K$, $M'L$ perpendicular to $OX$, $MP$.

Then the angle
$$LPM' = 90° - LM'P = LM'O = \theta;$$
and
$$x = OM = OK - MK = OK - LM'$$
$$= OM' \cos \theta - M'P \sin \theta,$$
or
$$x = x' \cos \theta - y' \sin \theta \qquad \ldots\ldots(1).$$
Similarly
$$y = MP = ML + LP = KM' + LP$$
$$= OM' \sin \theta + M'P \cos \theta,$$
or
$$y = x' \sin \theta + y' \cos \theta \qquad \ldots\ldots(2).$$

The substitution of these expressions for $x$ and $y$ in terms of $x'$, $y'$ enables us to transform any equation from one set of axes to the other.

**3.62  Examples.**

(i) *What does the equation*
$$x^2 - 3xy + 3y^2 + 7x - 18y + 32 = 0$$
*become if the origin is moved to the point* $(4, 5)$ *and the axes are turned through* $45°$?

First move the origin by writing $x' + 4$ for $x$ and $y' + 5$ for $y$. This gives

$(x' + 4)^2 - 3(x' + 4)(y' + 5) + 3(y' + 5)^2 + 7(x' + 4) - 18(y' + 5) + 32 = 0,$

which reduces to $\qquad x'^2 - 3x'y' + 3y'^2 + 1 = 0,$

or suppressing the dashes

$$x^2 - 3xy + 3y^2 + 1 = 0.$$

Next turn the axes through $45°$ by writing $\dfrac{x' - y'}{\sqrt{2}}$ for $x$ and $\dfrac{x' + y'}{\sqrt{2}}$ for $y$ in the last equation.

This becomes

$$(x' - y')^2 - 3(x'^2 - y'^2) + 3(x' + y')^2 + 2 = 0,$$
or $\qquad x'^2 + 4x'y' + 7y'^2 + 2 = 0,$

which is the required equation.

(ii) *Through what angle must the axes be turned in order to make the coefficient of xy zero in the equation*

$$x^2 + 2\sqrt{3}xy - y^2 = 4?$$

If we turn the axes through an angle $\theta$ the equation becomes
$(x'\cos\theta - y'\sin\theta)^2 + 2\sqrt{3}(x'\cos\theta - y'\sin\theta)(x'\sin\theta + y'\cos\theta)$
$\qquad\qquad\qquad\qquad - (x'\sin\theta + y'\cos\theta)^2 = 4.$

The coefficient of $x'y'$ in this equation is

$$- 2\sin\theta\cos\theta + 2\sqrt{3}(\cos^2\theta - \sin^2\theta) - 2\sin\theta\cos\theta,$$

and this vanishes if

$$\sin 2\theta = \sqrt{3}\cos 2\theta \text{ or } \tan 2\theta = \sqrt{3};$$

therefore $30°$ is the required angle.

### 3.63 Examples.

1. What does the equation

$$x^2 - y^2 - 4x - 6y - 6 = 0$$

become when the origin is moved to the point $(2, -3)$?

2. To what point must the origin be moved in order that the equation $\qquad 2x^2 - 3xy + 4y^2 + 10x - 19y + 23 = 0$

may become $\qquad 2x^2 - 3xy + 4y^2 = 1$?

3. Show that the equation

$$x^2 + y^2 = a^2$$

remains unaltered by any rotation of the axes.

**4.** What does the equation

$$x^2 + 2\sqrt{3}xy - y^2 = 2a^2$$

become when the axes are turned through 30°?

**5.** What does the equation

$$(y - x)^2 = 4\sqrt{2}a\,(x + y)$$

become when the axes are turned through 45°?

**6.** To what point must the origin be moved in order that the equation

$$x^2 + 4xy - 2y^2 + 10x - 4y = 0$$

may become

$$x^2 + 4xy - 2y^2 = 1,$$

and through what angle must the axes be turned in order to get rid of the term in $xy$?

**7.** Through what angle must the axes be turned to reduce the equation

$$x^2 - 2xy - y^2 = 1$$

to the form

$$xy = \text{const.?}$$

**8.** Show that, by changing the origin, the equation

$$2x^2 + 2y^2 + 7x + 5y - 13 = 0$$

can be transformed to     $8x^2 + 8y^2 = 89.$

**9.** Show that, by rotating the axes, the equation

$$8x^2 + 7xy - 3y^2 = 1$$

can be reduced to       $\sqrt{85}\,(x^2 - y^2) = 2.$

**10.** Show that, by rotating the axes, the equation

$$41x^2 + 24xy + 34y^2 = 75$$

can be reduced to         $2x^2 + y^2 = 3.$

**11.** Show that, by a change of origin and the directions of the co-ordinate axes, the equation

$$5x^2 + 2xy + 5y^2 - 14x - 22y + 27 = 0$$

can be transformed to      $3x^2 + 2y^2 = 1,$

or                       $2x^2 + 3y^2 = 1.$

# CHAPTER IV

# THE CIRCLE

**4.1 To find the equation of a circle when the centre and radius are given.** Let $r$ be the radius and $a$, $b$ the co-ordinates of the centre $C$; then, if $x$, $y$ are the co-ordinates of any point $P$ on the circle, by expressing the fact that $r$ is the distance of $P$ from $C$ in terms of their co-ordinates we get (1.3)

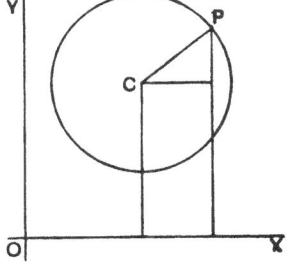

$$(x - a)^2 + (y - b)^2 = r^2 \quad \ldots\ldots(1).$$

This is the equation required and it can also be written

$$x^2 + y^2 - 2ax - 2by + a^2 + b^2 - r^2 = 0 \quad \ldots\ldots(2).$$

When the origin is at the centre of the circle the equation is

$$x^2 + y^2 = r^2 \quad \ldots\ldots(3).$$

**4.11 The equation $x^2 + y^2 + 2gx + 2fy + c = 0$ represents a circle.** This follows from the fact that the equation may be written

$$(x + g)^2 + (y + f)^2 = g^2 + f^2 - c,$$

expressing the fact that the square of the distance of $(x, y)$ from the fixed point $(-g, -f)$ is a constant number $g^2 + f^2 - c$. But there are three possibilities:

(i) when $g^2 + f^2 - c$ is positive, the locus is a real circle of centre $(-g, -f)$ and radius $\sqrt{(g^2 + f^2 - c)}$;

(ii) when $g^2 + f^2 - c = 0$, the locus $(x + g)^2 + (y + f)^2 = 0$ is called a point circle, the only real solution of the equation being $x = -g$, $y = -f$;

(iii) when $g^2 + f^2 - c$ is negative, the locus is an imaginary circle.

**4.12** In order that an equation may represent a circle it must be possible to arrange it in the form (1), that is to say the equation must be of the second degree, $x^2$ and $y^2$ must have the same coefficient and there must be no term in $xy$.

**4.13** Example. *Find the centre and radius of the circle*
$$2x^2 + 2y^2 - 14x + 10y + 19 = 0.$$

Dividing by 2 we get
$$x^2 + y^2 - 7x + 5y + \tfrac{19}{2} = 0,$$
or
$$(x - \tfrac{7}{2})^2 + (y + \tfrac{5}{2})^2 = \tfrac{49}{4} + \tfrac{25}{4} - \tfrac{19}{2} = 9.$$
Therefore the centre is $(\tfrac{7}{2}, -\tfrac{5}{2})$ and the radius is 3.

**4.14** Since the equation of a circle in general contains three independent constants, viz. $g$, $f$ and $c$ of 4.11, therefore one circle in general can be drawn to pass through any three arbitrarily chosen points. For on substituting the co-ordinates of the points in turn in the equation
$$x^2 + y^2 + 2gx + 2fy + c = 0$$
we get three equations of the first degree to find $g$, $f$ and $c$ and in general they have one and only one solution.

**4.15** Example. *Find the equation of the circle that passes through the points* $(0, 1)$, $(2, 3)$, $(-2, 5)$.

Let
$$x^2 + y^2 + 2gx + 2fy + c = 0 \qquad \ldots\ldots(1)$$
be the equation of the circle. By substituting the co-ordinates of the given points in turn in equation (1) we get
$$1 + 2f + c = 0,$$
$$13 + 4g + 6f + c = 0,$$
and
$$29 - 4g + 10f + c = 0;$$
from which we find that $f = -\tfrac{10}{3}$, $g = \tfrac{1}{3}$ and $c = \tfrac{17}{3}$.

Therefore the required equation is
$$3x^2 + 3y^2 + 2x - 20y + 17 = 0.$$

**4.16** Examples.

1. Find the radius and the co-ordinates of the centre of each of the following circles:

(i) $3x^2 + 3y^2 - 12x + 6y + 11 = 0$;

(ii) $(x - a)(x + a) + (y - b)(y + b) = 0$;

(iii) $2x^2 + 2y^2 + 16x - 4y + 33 = 0$;

(iv) $36x^2 + 36y^2 - 36x - 24y - 131 = 0$.

**2.** Find the equation of the circle that passes through the points $(1, 2), (2, 1), (0, 0)$.

**8.** Find the equation of the circle that passes through the points $(2, 3), (3, 2), (5, 1)$.

**4.** Find the equation of the circle that passes through the points $(2a, 0), (0, 2b), (a + b, a + b)$.

**5.** A circle has its centre on the line $x = 2y$ and passes through the points $(-1, 2), (3, -2)$. Find the co-ordinates of the centre and the equation of the circle.

**6.** Find the locus of the centre of a circle which touches the line $x \cos \alpha + y \sin \alpha = p$ and the circle $(x - a)^2 + (y - b)^2 = c^2$.

**4.2** Many of the properties of a circle that are proved in elementary pure geometry can also be easily proved by co-ordinate geometry. For example:

*The angle in a semicircle is a right angle.*

Take the origin at the centre of the circle, let the axis $OX$ cut it in $A, B$ and let $P$ be any point $(x, y)$ on the circle.

Then, if $a$ is the radius, the points $A, B$ are the points $(-a, 0), (a, 0)$ so that the $m$ of the line $AP$ is $y/(x + a)$, and the $m$ of the line $BP$ is $y/(x - a)$, and the condition of perpendicularity of the two lines, i.e. the condition $1 + mm' = 0$ of 2.52, is in this case

$$1 + \frac{y^2}{x^2 - a^2} = 0,$$

or $$x^2 + y^2 = a^2,$$

which is satisfied for every point $P$ on the circle.

**4.21** To find the equation of a circle having given the co-ordinates of the ends of a diameter. Let $x', y'$ and $x'', y''$ be the co-ordinates of the ends $A, B$ of a diameter. Let $x, y$ be the co-ordinates of any point $P$ on the circle.

Then since $APB$ is a semicircle the lines $PA, PB$ are at right angles, but $AP$ makes with $OX$ an angle $\tan^{-1} \dfrac{y - y'}{x - x'}$ and $PB$ makes with $OX$ an angle $\tan^{-1} \dfrac{y - y''}{x - x''}$ and

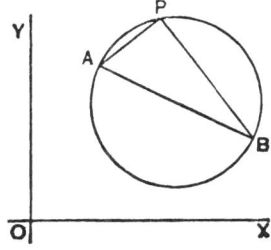

the condition of perpendicularity, i.e. the condition $1 + mm' = 0$ of 2·52, is in this case

$$1 + \frac{y - y'}{x - x'} \cdot \frac{y - y''}{x - x''} = 0,$$

or

$$(x - x')(x - x'') + (y - y')(y - y'') = 0.$$

This is therefore the locus of a point $P$ such that $APB$ is a right angle, i.e. it represents the circle on $AB$ as diameter.

**4.3   Segments of chords.**   *If through a point $P$ whose co-ordinates are $x'$, $y'$ a line be drawn cutting the circle $x^2 + y^2 = a^2$ in $Q$, $R$ then*
$$PQ.PR = x'^2 + y'^2 - a^2,$$
*so that this product is independent of the direction of the line $PQR$ and depends only on the radius of the circle and the distance of $P$ from its centre.*

Let the line $PQR$ make an angle $\theta$ with $OX$, then its equation is

$$\frac{x - x'}{\cos \theta} = \frac{y - y'}{\sin \theta} = r \quad (2.43).$$

It follows that, if $Q$ (or $R$) is at a distance $r$ from $P$, its co-ordinates are   $x' + r \cos \theta, \quad y' + r \sin \theta.$

But $Q$ and $R$ lie on the circle, so that

$$(x' + r \cos \theta)^2 + (y' + r \sin \theta)^2 = a^2,$$

or, arranging as a quadratic in $r$,

$$r^2 + 2r (x' \cos \theta + y' \sin \theta) + x'^2 + y'^2 - a^2 = 0 \quad \ldots\ldots(1).$$

Now the roots of this quadratic are the distances $PQ$, $PR$, and, by equating the product of the roots to the third term in the equation, we get   $PQ.PR = x'^2 + y'^2 - a^2.$

It must be noted that this result is to be interpreted algebraically, so that the product $PQ.PR$ is positive when $PQ$, $PR$ are in the same sense and negative when $PQ$, $PR$ are in opposite senses, i.e. positive when $P$ is outside and negative when inside the circle.

It follows that $x'^2 + y'^2 - a^2 > = $ or $< 0$, according as the point $(x', y')$ is outside, on the circumference, or inside the circle

$$x^2 + y^2 - a^2 = 0.$$

When $P$ is outside the circle, if $PT$ be a tangent to the circle, i.e. a position of the line $PQR$ for which $Q$ and $R$ coincide, then

$$PT^2 = PQ.PR = x'^2 + y'^2 - a^2 = OP^2 - OT^2,$$

thus proving that the tangent $PT$ is at right angles to the radius $OT$.

The expression $x'^2 + y'^2 - a^2$ is sometimes called **the power of the point** $(x', y')$ **with regard to the circle** $x^2 + y^2 - a^2 = 0$. It follows that, when the power is positive, the point lies outside the circle and the power is the square of the tangent from the point to the circle; and, when the power is negative, the point lies within the circle and the power (with sign changed) is the rectangle under the segments of chords through the point.

**4.31** The properties of the circle proved in the last article are purely geometrical and do not depend on the form of the equation to the circle. If we take the equation of the circle to be

$$x^2 + y^2 + 2gx + 2fy + c = 0,$$

we find for the quadratic in $r$ the equation

$$r^2 + 2r \{(x'+g)\cos\theta + (y'+f)\sin\theta\}$$
$$+ x'^2 + y'^2 + 2gx' + 2fy' + c = 0 \qquad \ldots\ldots(1),$$

so that the product $PQ.PR$ or the power of the point $(x', y')$ with regard to the circle is $x'^2 + y'^2 + 2gx' + 2fy' + c$;

and the point $(x', y')$ is outside, on the circumference, or inside the circle according as

$$x'^2 + y'^2 + 2gx' + 2fy' + c > = \text{ or } < 0.$$

The quadratic (1) in this article and the last may be used in other connections as will appear in subsequent articles.

**4.32** We remarked in 4.11 that $-g, -f$ are the co-ordinates of the centre of the circle $x^2 + y^2 + 2gx + 2fy + c = 0$,

and it follows from 4.31 that $c$ is the power of the origin, or the rectangle under the segments of chords through the origin; and the origin lies outside, on the circumference or inside the circle according as $c > = \text{ or } < 0$.

**4.33** To find the equation of the tangent at the point $(x', y')$ on the circle $x^2 + y^2 + 2gx + 2fy + c = 0$.

We have seen in 4.31 that the line

$$\frac{x - x'}{\cos\theta} = \frac{y - y'}{\sin\theta} = r \qquad \ldots\ldots(1)$$

through the point $P$ whose co-ordinates are $x'$, $y'$ meets the circle in points $Q$, $R$ whose distances from $P$ are the roots of the quadratic

$$r^2 + 2r\,\{(x'+g)\cos\theta + (y'+f)\sin\theta\}$$
$$+ x'^2 + y'^2 + 2gx' + 2fy' + c = 0 \qquad \ldots\ldots(2).$$

Let $(x', y')$ be on the circle, i.e. let $P$ coincide with $Q$ (fig. 4.3), then the last term in the quadratic is zero and one of its roots (viz. $PQ$) is zero. Then if the direction of the line be chosen so that $R$ also coincides with $Q$ and $P$ the line is the tangent to the circle at $P$, and the other root of the quadratic (viz. $PR$) is also zero. This requires that

$$(x'+g)\cos\theta + (y'+f)\sin\theta = 0 \qquad \ldots\ldots(3).$$

This equation gives $\theta$, i.e. the inclination to $OX$ of the tangent to the circle at $(x', y')$; and by eliminating $\theta$ between (1) and (3) we get the equation of the tangent in the form

$$(x-x')(x'+g) + (y-y')(y'+f) = 0,$$
or $\qquad xx' + yy' + gx + fy - x'^2 - y'^2 - gx' - fy' = 0.$

But since $(x', y')$ is on the circle we have

$$x'^2 + y'^2 + 2gx' + 2fy' + c = 0;$$

and by adding the last two equations we get

$$xx' + yy' + g(x+x') + f(y+y') + c = 0 \qquad \ldots\ldots(4)$$

which is the standard form of equation of the tangent.

In like manner we can show that the equation of the tangent at $(x', y')$ on the circle $x^2 + y^2 = a^2$ is

$$xx' + yy' = a^2 \qquad \ldots\ldots(5).$$

The student is advised to write out the proof as an exercise. A direct method of finding the equation of the tangent is given in the following article.

### 4.34 Equation of the tangent to a circle at a given point.

(i) Let the equation of the circle be

$$x^2 + y^2 = a^2 \qquad \ldots\ldots(1)$$

and $(x', y')$ the point of contact of the tangent.

By differentiating (1) and dividing by 2 we get

$$x + y\frac{dy}{dx} = 0,$$
or $\qquad \dfrac{dy}{dx} = -\dfrac{x}{y}.$

Therefore the gradient at $(x', y')$ is $-\dfrac{x'}{y'}$, and the equation of the tangent is
$$y - y' = -\frac{x'}{y'}(x - x') \quad (3.32).$$

This reduces to $\quad xx' + yy' = x'^2 + y'^2.$

But, since $(x', y')$ lies on the circle, therefore $x'^2 + y'^2 = a^2$, and the equation of the tangent is
$$xx' + yy' = a^2 \qquad \ldots\ldots(2).$$

(ii) Let the equation of the circle be
$$x^2 + y^2 + 2gx + 2fy + c = 0 \qquad \ldots\ldots(3).$$

Proceeding in the same way we have
$$x + y\frac{dy}{dx} + g + f\frac{dy}{dx} = 0,$$

or
$$\frac{dy}{dx} = -\frac{x+g}{y+f}.$$

Therefore the gradient at $(x', y')$ is $-\dfrac{x'+g}{y'+f}$ (compare **4.33** (3)), and the equation of the tangent at $(x', y')$ is
$$y - y' = -\frac{x'+g}{y'+f}(x - x') \quad (3.32).$$

This reduces to
$$xx' + yy' - x'^2 - y'^2 + g(x - x') + f(y - y') = 0 \quad \ldots(4).$$

But, since $(x', y')$ lies on the circle, therefore
$$x'^2 + y'^2 + 2gx' + 2fy' + c = 0 \qquad \ldots\ldots(5),$$

and by adding (5) and (4) we obtain the equation of the tangent at $(x', y')$ in the form
$$xx' + yy' + g(x + x') + f(y + y') + c = 0 \quad \ldots\ldots(6).$$

**4.35 Equation of the normal at a point on a circle.**

(i) When the equation of the circle is $x^2 + y^2 = a^2$, the tangent at $(x', y')$ is $xx' + yy' = a^2$, and the normal passes through $(x', y')$ and is perpendicular to the tangent, so that its equation is
$$y'(x - x') - x'(y - y') = 0 \quad (2.521),$$

or
$$y'x - x'y = 0.$$

This line passes through the origin which is the centre of the circle, thus proving that a tangent is at right angles to the radius drawn to the point of contact.

(ii) When the equation of the circle is

$$x^2 + y^2 + 2gx + 2fy + c = 0$$

the tangent at $(x', y')$ is

$$xx' + yy' + g(x + x') + f(y + y') + c = 0,$$

and the same argument as above shows that the normal is

$$(y' + f)(x - x') - (x' + g)(y - y') = 0,$$

or $\qquad (y' + f)x - (x' + g)y - fx' + gy' = 0.$

And it is easy to verify that this passes through the centre $(-g, -f)$ of the circle.

**4.36** To find the condition that the line $y = mx + c$ may touch the circle $x^2 + y^2 = a^2$. Eliminating $y$ between the two equations, we get $\qquad x^2 + (mx + c)^2 = a^2.$

This is a quadratic for the abscissae of the two points in which the given line cuts the circle, viz.

$$x^2(1 + m^2) + 2mcx + c^2 - a^2 = 0.$$

If the line touches the circle the two points of intersection coincide and the quadratic has equal roots. The condition for this is $\qquad m^2c^2 - (1 + m^2)(c^2 - a^2) = 0,$

which reduces to $\qquad c^2 = a^2(1 + m^2),$

or $\qquad c = \pm a\sqrt{(1 + m^2)}.$

We conclude that of all the parallel straight lines which have a given $m$, there are two which touch the circle, namely

$$y = mx + a\sqrt{(1 + m^2)} \text{ and } y = mx - a\sqrt{(1 + m^2)}.$$

**4.361** From 4.35 it follows that a line touches a circle if its perpendicular distance from the centre is equal to the radius. This gives another proof of the result of the last article, for the origin is the centre of the circle and the distance of $y = mx + c$ from the origin is $\pm c/\sqrt{(1 + m^2)}$, so that the required condition is $\pm c/\sqrt{(1 + m^2)} = a.$

**4.362  Two tangents can be drawn from any point to a circle.**
The line
$$y = mx + a\sqrt{(1 + m^2)}$$
touches the circle $x^2 + y^2 = a^2$, and it passes through a specified point $(h, k)$ if
$$k = mh + a\sqrt{(1 + m^2)},$$
i.e. if
$$m^2(h^2 - a^2) - 2mhk + k^2 - a^2 = 0.$$

This quadratic in $m$ gives the directions of the two tangents that pass through the point $(h, k)$.

They are real and distinct, coincident or imaginary according as
$$h^2 k^2 - (h^2 - a^2)(k^2 - a^2) > = \text{ or } < 0,$$
i.e. as
$$h^2 + k^2 - a^2 > = \text{ or } < 0,$$
i.e. according as $(h, k)$ is outside, upon or inside the circumference.

**4.4  Middle points of chords.** We have seen in **4.3** that the line
$$\frac{x - x'}{\cos \theta} = \frac{y - y'}{\sin \theta} = r \qquad \ldots\ldots(1)$$
which passes through a point $P$ whose co-ordinates are $x'$, $y'$, meets the circle $x^2 + y^2 = a^2$ in points $Q$, $R$ whose distances from $P$ are the roots of the quadratic
$$r^2 + 2r(x'\cos\theta + y'\sin\theta) + x'^2 + y'^2 - a^2 = 0 \quad \ldots\ldots(2).$$

Let $P$ be the middle point of the chord $QR$, then $PQ = -PR$, or the sum of the roots is zero.

Therefore $x'\cos\theta + y'\sin\theta = 0$  $\ldots\ldots(3)$.

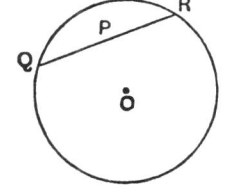

This shows that if $\theta$ is a given angle, i.e. if $QR$ is any one of a set of parallel chords, the middle point $P$ of the chord lies on a line $y = -x\cot\theta$. This is a line through the centre perpendicular to the chord (1).

Hence the locus of the middle points of a set of parallel chords of the circle is a straight line through the centre at right angles to the chords.

From (1) and (3) we may deduce the equation of a chord whose middle point has given co-ordinates $(x', y')$. By eliminating $\theta$ we get
$$x'(x - x') + y'(y - y') = 0,$$
or
$$xx' + yy' = x'^2 + y'^2,$$
as the required equation.

### 4.41 Examples.

(i) *Find the equations of the tangents to the circle*

$$x^2 + y^2 - 4x - 6y + 9 = 0$$

*which are parallel to the line* $4x + 3y = 0$ *and the co-ordinates of their points of contact.*

Let $4x + 3y = c$ be the equation of one of the tangents. It meets the circle where

$$x^2 + \tfrac{1}{9}(c - 4x)^2 - 4x - 2(c - 4x) + 9 = 0,$$

which reduces to

$$25x^2 + 2(18 - 4c)x + c^2 - 18c + 81 = 0 \qquad \ldots\ldots(1).$$

Since the line touches the circle this equation, which gives the abscissae of the intersections, must have equal roots.

Therefore $(18 - 4c)^2 = 25(c^2 - 18c + 81)$,

or $18 - 4c = \pm 5(c - 9)$,

i.e. $c = 7$ or $27$.

Hence the equations of the tangents are

$$4x + 3y = 7 \text{ and } 4x + 3y = 27.$$

To find the points of contact, the abscissae are given by equation (1) when we give $c$ the values 7 and 27; but the equation (1) then has equal roots and either root is half the sum of the roots, i.e. either root is $-(18 - 4c)/25$.

Hence, when we take $c = 7$, we have a tangent $4x + 3y = 7$ which touches the circle at the point

$$x = -\frac{18 - 4c}{25} = \tfrac{2}{5}, \ y = \tfrac{9}{5};$$

and, when we take $c = 27$, we have a tangent $4x + 3y = 27$ which touches the circle at the point

$$x = -\frac{18 - 4c}{25} = 1\tfrac{3}{5}, \ y = 2\tfrac{1}{5}.$$

*Alternatively*, to find the values of $c$. The equation of the circle is

$$(x - 2)^2 + (y - 3)^2 = 4,$$

so that its centre is (2, 3) and its radius 2. And the distance of a tangent $4x + 3y = c$ from the centre must be equal to the radius, so that

$$\pm \frac{4 \cdot 2 + 3 \cdot 3 - c}{5} = 2 \quad (2.62),$$

giving $c = 7$ and 27 as before.

(ii) *Find the equation of a circle which passes through the point* $(3, 2)$ *and touches the line* $2x + 3y = 5$ *at the point* $(1, 1)$.

Let $x^2 + y^2 + 2gx + 2fy + c = 0$ be the equation of the circle.

By 4.33 or 4.34 the tangent at the point $(1, 1)$ is

$$x + y + g(x + 1) + f(y + 1) + c = 0,$$

and by comparing this equation with $2x + 3y = 5$, we have

$$\frac{1+g}{2} = \frac{1+f}{3} = \frac{g+f+c}{-5} \qquad \dots\dots(1).$$

Again, since the point $(3, 2)$ lies on the circle we have

$$13 + 6g + 4f + c = 0 \qquad \dots\dots(2).$$

Solving (1) and (2) we find that

$$f = -\tfrac{29}{14}, \; g = -\tfrac{12}{7} \text{ and } c = \tfrac{39}{7},$$

so that the circle is

$$7x^2 + 7y^2 - 24x - 29y + 39 = 0.$$

*Alternatively*, let the centre be the point $(a, b)$; then $a$, $b$ can be found from the equations

$$(a-3)^2 + (b-2)^2 = (a-1)^2 + (b-1)^2 = \frac{(2a+3b-5)^2}{13} = (\text{radius})^2;$$

since these equations express the fact that the radius is the distance of $(a, b)$ from $(3, 2)$ and from $(1, 1)$ and is also the perpendicular distance of $(a, b)$ from the line $2x + 3y = 5$.

The solution of these equations gives $a = \tfrac{12}{7}$, $b = \tfrac{29}{14}$ and radius $= \dfrac{5\sqrt{13}}{14}$.

### 4.42 Examples.

1. Without drawing a figure determine whether the points $(-1, 2)$, $(0, 0)$, $(3, -4)$ lie outside, on the circumference, or inside the circle

$$x^2 + y^2 - 5x + 2y - 5 = 0.$$

2. Find the points of intersection of the line $3x + 2y = 12$ and the circle $x^2 + y^2 = 13$, and find for what values of $c$ the line $3x + 2y = c$ touches the circle.

3. Prove that the line $3x + 2y = 30$ touches the circle

$$x^2 + y^2 - 10x - 2y + 13 = 0,$$

and find the co-ordinates of the point of contact.

4. For what values of $m$ does the line $y = mx$ touch the circle

$$x^2 + y^2 - 6x - 2y + 8 = 0?$$

5. Prove that the circle $(x-a)^2 + (y-a)^2 = a^2$ touches the co-ordinate axes.

6. Show that two circles can be drawn to pass through the point $(1, 2)$ and touch the co-ordinate axes, and find their equations.

7. Find the length of the tangent from the point $(7, 4)$ to the circle
$$x^2 + y^2 - 4x - 6y + 12 = 0.$$

8. Find the equations of the tangents to the circle
$$x^2 + y^2 - 4x - 3y + 5 = 0$$
that are parallel to the line $y + x = 0.$

9. Find the equations of the tangents to the circle
$$x^2 + y^2 - 7x - 5y + 18 = 0,$$
at the points $(4, 3)$ and $(3, 2)$, showing that they are parallel.

10. Find the equations of the tangent and normal to the circle
$$x^2 + y^2 - 6x + 4y - 12 = 0$$
at the point $(6, 2)$.

11. Prove that the line $x + y = 1$ touches the circle
$$x^2 + y^2 - 8x - 6y + 7 = 0,$$
and find the equations of the parallel and perpendicular tangents.

12. Find the equation of the tangent at the origin to the circle
$$x^2 + y^2 + 2gx + 2fy = 0.$$

13. Prove that the line $x \cos \alpha + y \sin \alpha = p$ touches the circle
$$(x-a)^2 + (y-b)^2 = r^2,$$
if $\qquad r = \pm (p - a \cos \alpha - b \sin \alpha).$

14. Find the points of contact of the tangents to the circle
$$x^2 + y^2 = 25$$
that pass through the point $(7, 1)$ and write down the equations of the tangents.

15. Prove that the tangent to the circle $x^2 + y^2 = 5$ at the point $(1, -2)$ also touches the circle
$$x^2 + y^2 - 8x + 6y + 20 = 0,$$
and find the co-ordinates of the point of contact.

16. Find the equations of the circles that touch the lines
$$y = 0, \quad y = 4, \quad 2x + y = 2.$$

**17.** Find the co-ordinates of the middle point of the chord

$$x + 7y = 25$$

of the circle
$$x^2 + y^2 = 25.$$

**18.** Find the equation of the chord of the circle

$$x^2 + y^2 - 6x - 4y - 23 = 0,$$

which has the point $(4, 1)$ as its middle point.

**19.** Prove that the circle

$$x^2 + y^2 - 6x - 4y + 9 = 0$$

subtends an angle $\tan^{-1} \tfrac{12}{5}$ at the origin.

**20.** Find the condition that the line

$$lx + my + n = 0$$

should touch the circle
$$(x - a)^2 + (y - b)^2 = r^2.$$

**21.** Verify that the perpendicular bisector of the chord joining two points $(x_1, y_1)$, $(x_2, y_2)$ on the circle

$$x^2 + y^2 + 2gx + 2fy + c = 0$$

passes through the centre.

**4.5  Chord of contact of tangents drawn from a given point to a circle.** Let $x'$, $y'$ be the co-ordinates of a point $P$ outside the circle $x^2 + y^2 = a^2$.

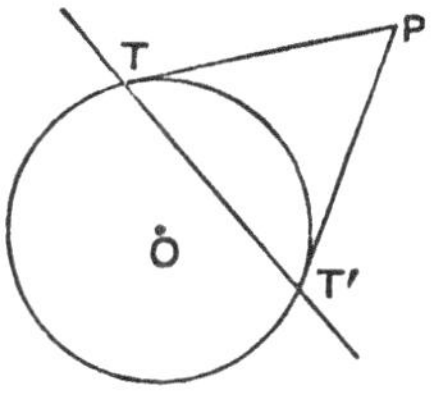

Let the tangents from $P$ touch the circle at points $T$, $T'$ whose co-ordinates are $h$, $k$ and $h'$, $k'$. Then the equations of $TP$ and $T'P$ are $xh + yk = a^2$

and $\qquad xh' + yk' = a^2$  (**4.33 or 4.34**).

But these lines both pass through $P$, therefore
$$x'h + y'k = a^2 \quad \text{and} \quad x'h' + y'k' = a^2 \qquad \ldots\ldots(1).$$

Consider now the equation

$$xx' + yy' = a^2 \qquad \ldots\ldots(2).$$

Being of the first degree it represents a straight line, and equations (1) show that it is satisfied by $x = h$, $y = k$ and by $x = h'$, $y = k'$, therefore it is the equation of the line $TT'$.

Thus it appears that when the point $(x', y')$ lies on the circle, the equation $xx' + yy' = a^2$ represents the tangent at the point; but when the point lies outside the circle the same equation represents the chord of contact of the tangents drawn from the point.

We note that since the equation of the line $OP$ is $\dfrac{x}{x'} = \dfrac{y}{y'}$, therefore $OP$ is at right angles to $TT'$.

### 4.51  Pole and polar.

DEF. *The polar of a point with regard to a circle (or conic) is the locus of the points of intersection of the pairs of tangents drawn from the ends of chords that pass through the point and in relation to its polar the point is called the pole.*

**To find the equation of the polar of the point $(x', y')$ with regard to the circle $x^2 + y^2 = a^2$.**

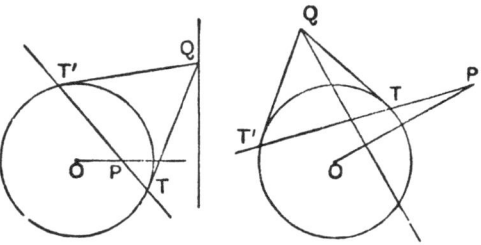

Let $P$ be the point $(x', y')$ which may be inside or outside the circle.

Let *any* line through $P$ cut the circle in $T$, $T'$ and let the tangents at $T$, $T'$ intersect in $Q$. Then $Q$ is a point on the polar of $P$.

Let $x''$, $y''$ be the co-ordinates of $Q$, then by **4.5** the equation of $TT'$ is
$$xx'' + yy'' = a^2;$$

but $TT'$ passes through $P$ whose co-ordinates are $x'$, $y'$, therefore

$$x'x'' + y'y'' = a^2 \qquad \ldots\ldots(1).$$

Now writing current co-ordinates for $x''$, $y''$ we have for the locus of $Q$ the line $$xx' + yy' = a^2 \qquad \ldots\ldots(2).$$

This is therefore the equation of the polar of $P$.

Hence the polar of a point with regard to a circle is a straight line, and by comparing with 4.5 we see that when the point $P$ lies outside the circle its polar is the chord of contact of tangents drawn from $P$ to the circle.

Also the form of the equation shows that the polar of $P$ is at right angles to the line joining $P$ to the centre.

**4.52  Construction of the polar of a given point.** The polar of a point $P$ or $(x', y')$ is $xx' + yy' = a^2$.

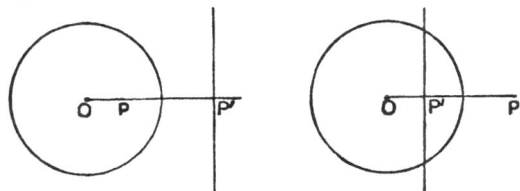

The perpendicular distance of this line from the centre $O$ of the circle is $\dfrac{a^2}{\sqrt{(x'^2 + y'^2)}}$, i.e. $a^2/OP$. Hence the polar of $P$ may be constructed by measuring along $OP$ a length $OP' = a^2/OP$ and drawing through $P'$ a perpendicular to $OP$.

**4.53**  If $P$ is on the circle then, since $OP \cdot OP' = a^2$, $P'$ is also on the circle, and the polar of a point on the circle being at right angles to $OP'$ is the tangent at the point. This is in accord with the fact that the equation of the polar of *any* point $(x', y')$ is $xx' + yy' = a^2$ whether the point be inside or outside the circle, and *if* the point $(x', y')$ lies on the circumference then this equation represents the tangent at the point.

**4.54  If the polar of P passes through Q, then the polar of Q passes through P.** Let $x'$, $y'$ and $x''$, $y''$ be the co-ordinates of $P$ and $Q$.

The polar of $P$ is $$xx' + yy' = a^2$$ and passes through $Q$ if $$x'x'' + y'y'' = a^2 \qquad \ldots\ldots(1).$$

But this relation also shows that the point $(x', y')$ lies on the line

$$xx'' + yy'' = a^2,$$

or that $P$ lies on the polar of $Q$.

**Cor.** If the polars of $P$ and $Q$ intersect in $R$, then $R$ is the pole of $PQ$, for by the foregoing theorems the polar of $R$ passes through both $P$ and $Q$. The triangle $PQR$ is said to be *self-conjugate* or *self-polar*.

#### 4.55 Conjugate points and lines.

**Def.** *A pair of points are conjugate with regard to a circle (or conic) when each lies on the polar of the other: and a pair of lines are conjugate with regard to a circle when each passes through the pole of the other.*

The points $(x', y')$, $(x'', y'')$ are conjugate with regard to the circle

$$x^2 + y^2 = a^2 \quad \text{if} \quad x'x'' + y'y'' = a^2$$

as in 4.54 (1).

*To find the condition that the lines*

$$lx + my + n = 0, \quad l'x + m'y + n' = 0$$

*may be conjugate with regard to the circle* $x^2 + y^2 = a^2$.

Let $(x', y')$ be the pole of

$$lx + my + n = 0,$$

then this equation which represents the polar of $(x', y')$ must be the same as

$$xx' + yy' - a^2 = 0,$$

so that by comparison $\quad \dfrac{x'}{l} = \dfrac{y'}{m} = -\dfrac{a^2}{n}.$

Hence the line $l'x + m'y + n' = 0$ passes through the pole $(x', y')$ of

$$lx + my + n = 0,$$

if $\qquad\qquad l'x' + m'y' + n' = 0,$

or if $\qquad\qquad -\dfrac{a^2 ll'}{n} - \dfrac{a^2 mm'}{n} + n' = 0,$

or if $\qquad\qquad ll' + mm' - \dfrac{nn'}{a^2} = 0,$

which is the required condition.

**4.56** We leave it to the student following the arguments of 4.5 and 4.51 to prove that the polar of the point $(x', y')$ with regard to the circle

$$x^2 + y^2 + 2gx + 2fy + c = 0$$

is $\qquad\qquad xx' + yy' + g(x + x') + f(y + y') + c = 0.$

**4.57 Examples.**

1. Write down the equations of the polars of the following points with regard to the circle $x^2 + y^2 = 6$:

(i) $(2, -1)$;          (ii) $(3, 4)$;

and deduce the co-ordinates of the pole of the line joining the points.

2. Find the poles of the following lines with regard to the circle $x^2 + y^2 = 9$:    (i) $3x + 4y = 7$;    (ii) $5x - y + 6 = 0$,

and verify that the polar of their point of intersection is the line joining their poles.

3. Show that the points $(4, -2)$, $(3, -6)$ are conjugate with regard to the circle
$$x^2 + y^2 = 24.$$

4. Prove that the lines
$$5x + 3y = 40, \quad 7x - 5y = 10$$
are conjugate with regard to the circle
$$x^2 + y^2 = 20.$$

5. Find the polar of the point $(5, 4)$ with regard to the circle
$$x^2 + y^2 - 4x - 3y - 8 = 0.$$

6. Find the pole of $lx + my + n = 0$ with regard to

(i) $x^2 + y^2 = a^2$;

(ii) $x^2 + y^2 + 2gx + 2fy + c = 0$.

7. Prove that if two lines at right angles are conjugate with regard to a circle one of them must pass through the centre.

8. Prove that, if the chords of contact of pairs of tangents to a circle from $P$ and $Q$ intersect in $R$, then the line joining $R$ to the centre is perpendicular to $PQ$.

## SYSTEMS OF CIRCLES

### 4.6 Condition that two circles may cut orthogonally.

Let $$x^2 + y^2 + 2gx + 2fy + c = 0$$

and $$x^2 + y^2 + 2g'x + 2f'y + c' = 0$$

be the equations of the circles. The circles are said to cut orthogonally if the tangents at a point of intersection are at right angles, and this implies that the radii drawn to a point of intersection are at right angles, or that the square on the line joining the centres is equal to the sum of the squares on the radii.

Hence the required condition is

$$(g - g')^2 + (f - f')^2 = g^2 + f^2 - c + g'^2 + f'^2 - c',$$

or $$2gg' + 2ff' - c - c' = 0.$$

### 4.7 Radical axis of two circles.

DEF. *The radical axis of two circles is the locus of a point from which the tangents to the circles are of equal length, or of a point which has the same power with regard to the circles.*

Let $$x^2 + y^2 + 2gx + 2fy + c = 0 \qquad \ldots\ldots(1)$$

and $$x^2 + y^2 + 2g'x + 2f'y + c' = 0 \qquad \ldots\ldots(2)$$

be the equations of the circles.

Equating the squares of the tangents from the point $(x, y)$ to the circles (4.3), we have

$$x^2 + y^2 + 2gx + 2fy + c = x^2 + y^2 + 2g'x + 2f'y + c' \qquad \ldots\ldots(3),$$

or $$2(g - g')x + 2(f - f')y + c - c' = 0 \qquad \ldots\ldots(4).$$

This is the equation of the radical axis of the two circles, and being of the first degree it represents a straight line. Moreover, if the coordinates of a point satisfy (1) and (2) they clearly also satisfy (3), so that the radical axis passes through the common points of the circles.

Again the line joining the centres $(-g, -f)$, $(-g', -f')$ of the circles has equation

$$\frac{x + g}{g - g'} = \frac{y + f}{f - f'},$$

and from (4) it appears that the radical axis is at right angles to the line of centres of the circles.

COR. If two circles touch one another it follows from the definition that the common tangent to the circles is their radical axis.

**4.71 Imaginary points and lines.** In elementary pure geometry we are accustomed to deal only with real points and lines, thus we say that a straight line meets a circle in two points, or in one point (tangency) or does not meet it at all. But in analytical geometry the equation of the line (first degree) and the equation of the circle (second degree) when solved simultaneously have two solutions, real and distinct, or real and equal, or imaginary, and we are thus led to say that any straight line meets any circle in two points which are either real and distinct, or coincident, or imaginary.

In the same way we can say that from *any* point two tangents can be drawn to a circle and that they are real, coincident or imaginary according as the point is outside, upon or inside the circle; and even when the tangents are imaginary they have a real chord of contact, namely the polar of the point (4.5, 4.51).

Again, in reference to the last article we see that the radical axis of two circles might be defined as the straight line through their points of intersection whether they intersect in real points or not.

**4.72** In the figures below $PP'$ is the radical axis of the two circles, and if we confined ourselves to *real* tangents a doubt would arise as to whether the part $AB$ of the line $PP'$ which falls within the circles belongs to the radical axis since no real tangents can be drawn from

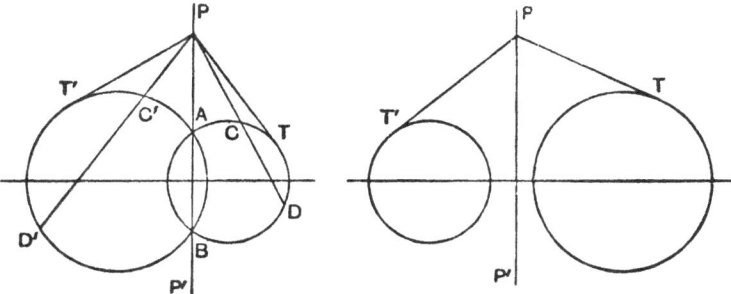

points on $AB$. But the conception of tangents real or imaginary from every point removes the doubt, and no such question arises if we take the alternative form of the definition in terms of "power of the point" instead of "length of tangent."

We note that when the circles cut in real points $A$, $B$ we have a simple verification that the line $AB$ is the locus required, because for any point $P$ on $AB$, or $AB$ produced,

$$PC.PD = PA.PB = PC'.PD'$$

so that the powers of $P$ with regard to the circles are the same.

**4.73** **The radical axes of three circles taken in pairs meet in a point.** Let

$$x^2 + y^2 + 2gx + 2fy + c = 0 \qquad \ldots\ldots(1),$$
$$x^2 + y^2 + 2g'x + 2f'y + c' = 0 \qquad \ldots\ldots(2),$$
$$x^2 + y^2 + 2g''x + 2f''y + c'' = 0 \qquad \ldots\ldots(3),$$

be the equations of the circles.

The equation of the radical axis of (2) and (3) is

$$2(g' - g'')x + 2(f' - f'')y + c' - c'' = 0 \qquad \ldots\ldots(4),$$

of (3) and (1)

$$2(g'' - g)x + 2(f'' - f)y + c'' - c = 0 \qquad \ldots\ldots(5),$$

and of (1) and (2)

$$2(g - g')x + 2(f - f')y + c - c' = 0 \qquad \ldots\ldots(6).$$

Now the sum of the equations (4), (5) and (6) is an identity, therefore the three lines have a common point (2.65). This point is called the **radical centre** of the three circles.

### 4.74 Coaxal circles.

DEF. *Circles are said to be coaxal when every pair of the circles have the same radical axis.*

It follows that the circles must have their centres collinear, for if $C_1$, $C_2$, $C_3$ ... are the centres then $C_1C_2$ is perpendicular to the radical axis and so are $C_1C_3$, $C_1C_4$ ..., therefore $C_1$, $C_2$, $C_3$ ... lie in the same straight line.

Take the radical axis for axis of $y$ and the line of centres for axis of $x$, then since the $y$ co-ordinate of the centre of each circle is zero the equations of any two are of the form

$$x^2 + y^2 + 2gx + c = 0$$

and

$$x^2 + y^2 + 2g'x + c' = 0,$$

so that their radical axis is

$$2(g - g')x + c - c' = 0.$$

But the radical axis is $x = 0$, therefore $c = c'$.

Hence the equation of any one of the coaxal circles is

$$x^2 + y^2 + 2gx + c = 0 \qquad \ldots\ldots(1),$$

where $c$ is the same for *all* the circles, but $g$ differs for the different members of the family.

**4.75** Families of coaxal circles are of two kinds according as they do or do not intersect in real points.

From 4.74 (1) we see that the coaxal circles *all* meet the radical axis $x = 0$ where

$$y^2 + c = 0,$$

i.e.

$$y = \pm \sqrt{(-c)}.$$

Hence if $c$ is negative the circles all cut the radical axis in the same two real points; but if $c$ is positive the circles do not cut the radical axis in real points, but the radical axis contains their intersections (real or imaginary) (4.7), therefore they have no real intersections.

Referring again to 4.74 (1), the centre of any circle of the coaxal family is $(-g, 0)$ and its radius is $\sqrt{(g^2 - c)}$. When $c$ is negative $g$ may be given any positive or negative numerical value and there is a corresponding real circle belonging to the family.

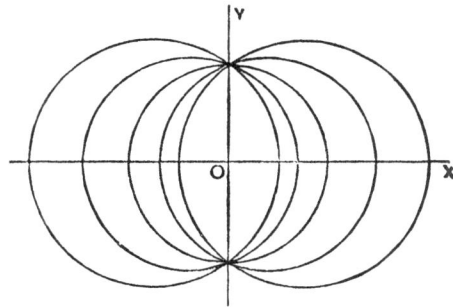

But when $c$ is positive $g^2$ must not be less than $c$ for the radius of the circle to be real and there are two circles of zero radius belonging to the family corresponding to $g = \pm \sqrt{c}$, these point circles are called the *limiting points* of the coaxal family. They are the points $L$, $L'$ in the figure, their co-ordinates being $\sqrt{c}, 0$ and $-\sqrt{c}, 0$.

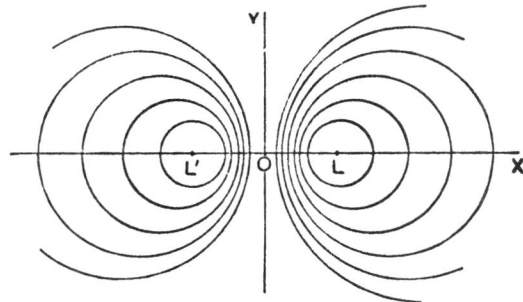

**4.76**   The polar of either limiting point with regard to any circle of a coaxal family passes through the other limiting point.   For if $L'$ be the point $(-\sqrt{c}, 0)$ its polar with regard to the circle

$$x^2 + y^2 + 2gx + c = 0,$$

is          $-x\sqrt{c} + y\,(0) + g\,(x - \sqrt{c}) + c = 0$    (4.56),

or                  $(g - \sqrt{c})\,(x - \sqrt{c}) = 0,$

i.e. the line $x = \sqrt{c}$, which passes through $L$.

**6**

**4.77   Orthogonal coaxal families.** Consider the equation

$$x^2 + y^2 + 2fy - c = 0 \qquad \ldots\ldots(1),$$

where $c$ is a positive constant. It represents a circle with its centre on the axis of $y$ and cutting the axis of $x$ in the fixed points $x = \pm \sqrt{c}$; so that for different values of $f$ it represents a coaxal family of circles which all intersect in the limiting points $(\sqrt{c}, 0)$, $(-\sqrt{c}, 0)$ of the coaxal family

$$x^2 + y^2 + 2gx + c = 0 \qquad \ldots\ldots(2).$$

It is easily seen that equations (1) and (2) satisfy the condition of 4.6.

It follows that two families of coaxal circles can be constructed such that the limiting points of one family are the points of intersection of the other, and every member of one family is cut orthogonally by every member of the other.

**4.771   Example.** *Find the limiting points of the circles*

$$x^2 + y^2 + 6x + 10 = 0, \quad x^2 + y^2 + 14x - 8y + 2 = 0.$$

The limiting points lie on the line of centres and any circle through them cuts the given circles orthogonally, so the limiting points may be found as the intersections of the line of centres and any circle orthogonal to the two given circles.

For simplicity let us take the circle that passes through the origin and is orthogonal to the two given circles, and let its equation be

$$x^2 + y^2 + 2gx + 2fy = 0.$$

The conditions of orthogonality (4.6) are

$$6g - 10 = 0 \quad \text{and} \quad 14g - 8f - 2 = 0,$$

whence $g = \tfrac{5}{3}, f = \tfrac{8}{3}$, so that

$$3x^2 + 3y^2 + 10x + 16y = 0 \qquad \ldots\ldots(1)$$

is a circle orthogonal to the given circles.

Again the centres of the given circles are $(-3, 0)$, $(-7, 4)$ and the equation of the line joining these points is

$$x + y + 3 = 0 \qquad \ldots\ldots(2).$$

Solving the equations (1) and (2) we find that the limiting points are

$$(-1 + \tfrac{1}{3}\sqrt{2}, \; -2 - \tfrac{1}{3}\sqrt{2}) \quad \text{and} \quad (-1 - \tfrac{1}{3}\sqrt{2}, \; -2 + \tfrac{1}{3}\sqrt{2}).$$

**4.78   Equation of a circle coaxal with two given circles.** Let

$$S \equiv x^2 + y^2 + 2gx + 2fy + c = 0 \qquad \ldots\ldots(1)$$

and

$$S' \equiv x^2 + y^2 + 2g'x + 2f'y + c' = 0 \qquad \ldots\ldots(2)$$

be the equations of the two circles.

Then $S - kS' = 0$ is also the equation of a circle, since it contains no term in $xy$ and its $x^2$ and $y^2$ terms have the same coefficients, and it is satisfied by the points common to $S = 0$ and $S' = 0$, so that it belongs to the same coaxal system.

Hence for different values of $k$ the equation $S - kS' = 0$ represents all the circles coaxal with $S = 0$ and $S' = 0$.

It follows that through any point in the plane not lying on either circle there can be drawn one and only one circle coaxal with two given circles; for by substituting the co-ordinates of the given point in the equation $S - kS' = 0$, we find one and only one value of $k$, and thus one and only one circle $S - kS' = 0$ which passes through the given point and is coaxal with $S = 0$ and $S' = 0$.

**4.79  Circles having a given radical axis.**  Let

$$S \equiv x^2 + y^2 + 2gx + 2fy + c = 0 \qquad \ldots\ldots(1)$$

be the equation of a given circle, and

$$L \equiv lx + my + n = 0 \qquad \ldots\ldots(2)$$

be the equation of a given line.

To find the equations of circles coaxal with (1) and having (2) for radical axis, we consider the equation

$$S - kL = 0 \qquad \ldots\ldots(3).$$

It is clearly the equation of a circle and it passes through the points common to $S = 0$ and $L = 0$, therefore for different values of $k$ it represents the family of coaxal circles required.

**4.8  Examples.**

1. Find the equation of a circle which passes through the points $(2, -1)$ and $(1, -2)$ and cuts orthogonally the circle

$$x^2 + y^2 - 2x + 3y - 5 = 0.$$

2. Find the equation of a circle which cuts orthogonally the three circles   $x^2 + y^2 + 4x - 5y + 6 = 0$,   $x^2 + y^2 + 5x - 6y + 7 = 0$,
and   $x^2 + y^2 - x - y - 1 = 0$.

3. Find the equation of a circle which cuts orthogonally the two circles   $x^2 + y^2 - 2x - 2y + 1 = 0$   and   $x^2 + y^2 - 3x + 6y - 2 = 0$,
and passes through the point $(-3, 2)$.

4. Write down the equations of the radical axes of the following pairs of circles:

(i) $x^2 + y^2 - 4x + 5y - 2 = 0$,   $x^2 + y^2 - 5x + 6y = 0$;
(ii) $x^2 + y^2 + 3x - 2y + 1 = 0$,   $x^2 + y^2 - 3x + 5y + 2 = 0$;
(iii) $x^2 + y^2 + 2gx + c = 0$,   $x^2 + y^2 + 2fy + c = 0$.

**5.** Find the equation of a circle coaxal with

$$x^2 + y^2 - 2x + 3y - 1 = 0 \quad \text{and} \quad x^2 + y^2 + 3x - 2y - 1 = 0$$

and passing through the point $(2, 1)$.

**6.** Find the co-ordinates of the point from which the tangents to the three circles

$$x^2 + y^2 - 4x - 4y + 7 = 0, \quad x^2 + y^2 + 4x + 3 = 0$$

and                              $$x^2 + y^2 + 2y = 0$$

are of equal length.

**7.** Find the limiting points of the circles

$$x^2 + y^2 + 2y - 4 = 0 \quad \text{and} \quad x^2 + y^2 + 2x + 2y - 10 = 0.$$

**8.** Prove that if a point moves so that the tangent from it to the circle
$$x^2 + y^2 + 4x - 5y + 6 = 0$$

is double the length of the tangent to the circle

$$x^2 + y^2 - 4 = 0$$

its locus is the circle $3x^2 + 3y^2 - 4x + 5y - 22 = 0.$

**9.** Prove that the locus of a point such that the lengths of the tangents from it to two given circles are in a constant ratio is a circle coaxal with the given circles.

**10.** Find the equations of the two circles coaxal with

$$x^2 + y^2 - 8x + 10y + 2 = 0 \quad \text{and} \quad x^2 + y^2 - 3x + 5y - 1 = 0$$

that touch the line           $$2x + y - 3 = 0.$$

**11.** Find the centre and radius of the circle which cuts orthogonally the three circles

$$x^2 + y^2 - 6x - 4y + 12 = 0, \quad x^2 + y^2 + 2x + 2y + 1 = 0$$

and                        $$x^2 + y^2 + 4x - 2y + 4 = 0.$$

**12.** The line $x + 3y + 2 = 0$ is the radical axis of a family of coaxal circles of which one circle is

$$x^2 + y^2 + 2x + 5y - 1 = 0.$$

Find the equation of the member of the family that passes through the point $(-3, 1)$.

### 4.9 Miscellaneous examples.

**1.** Find the locus of a point which moves so that the sum of the squares of its distances from the sides of an equilateral triangle is constant.

**2.** Find the locus of a point which moves so that the sum of the squares of its distances from $n$ fixed points is constant.

**3.** Find the locus of a point at which two given circles subtend equal angles.

**4.** A circle passes through the four points $(a, 0)$, $(b, 0)$, $(0, c)$, $(0, d)$. By what relation are $a, b, c, d$ connected? Find the equation of the circle and show that the tangent at the point $(a, c + d)$ is

$$(a - b)(x - a) + (c + d)(y - c - d) = 0.$$

**5.** Write down the equations of the tangents to the circles

$$x^2 + y^2 = 2ax, \quad x^2 + y^2 = 2by$$

at their points of intersection and verify that they cut at right angles.

**6.** Find the equation of the tangent to the circle $x^2 + y^2 = a^2$ at the point $(a \cos \theta, \ a \sin \theta)$ and show that the length of the tangent intercepted by the lines $x^2 - y^2 = 0$ is $\pm 2a \sec 2\theta$.

**7.** $A$ and $B$ are two fixed points $(c, 0)$, $(- c, 0)$ and $P$ moves so that $PA = k \cdot PB$. Find the locus of $P$ and prove that it is cut orthogonally by any circle through $A$ and $B$.

**8.** Show that the common chord of the circles

$$x^2 + y^2 - 6x - 4y + 9 = 0 \quad \text{and} \quad x^2 + y^2 - 8x - 6y + 23 = 0$$

is a diameter of the latter circle and find the angle at which the circles cut.

**9.** Prove analytically that the tangents to a circle at the ends of a chord are equally inclined to the chord.

**10.** Prove that for different values of $a$ the equation

$$x^2 + y^2 - 2ax \operatorname{cosec} \alpha + a^2 \cot^2 \alpha = 0$$

represents a family of circles touching the lines $y = \pm x \tan \alpha$.

Prove also that the locus of the poles of the line $lx + my = 0$ with regard to the circles is the line

$$mx \sin^2 \alpha + ly \cos^2 \alpha = 0.$$

**11.** Find the co-ordinates of the middle point of the chord $lx + my = 1$ of the circle

$$x^2 + y^2 + 2gx + 2fy + c = 0.$$

**12.** Prove that the points of intersection of the line $lx + my = 1$ and the circle
$$x^2 + y^2 + 2gx + 2fy + c = 0$$

subtend a right angle at the origin if

$$c\,(l^2 + m^2) + 2gl + 2fm + 2 = 0.$$

**13.** Prove that the equation of the circle having for diameter the portion of the line $x \cos\alpha + y \sin\alpha = p$ intercepted by the circle $x^2 + y^2 = a^2$ is $\qquad x^2 + y^2 - 2p\,(x \cos\alpha + y \sin\alpha - p) - a^2 = 0.$

**14.** Prove that if a chord of the circle $x^2 + y^2 = a^2$ subtends a right angle at a fixed point $(x', y')$ the locus of the middle point of the chord is $\qquad 2\,(x^2 + y^2 - xx' - yy') + x'^2 + y'^2 - a^2 = 0.$

**15.** Prove that the equation of any tangent to the circle

$$(x - a)^2 + (y - b)^2 = r^2$$

may be written in the form

$$(x - a) \cos\theta + (y - b) \sin\theta = r.$$

Deduce that the equation of the tangents from $(x', y')$ to the circle is
$$r^2\,\{(x - x')^2 + (y - y')^2\} = \{(x - a)\,(y' - b) - (y - b)\,(x' - a)\}^2.$$

**16.** Prove that the distances of two points from the centre of a circle are proportional to the distance of each point from the polar of the other.

**17.** Prove that the tangents to the circles of a coaxal system drawn from a limiting point are bisected by the radical axis.

**18.** Show that a common tangent to two circles is bisected by their radical axis and subtends a right angle at either limiting point.

**19.** Prove that if a point moves so that the difference of the squares of the tangents from it to two given circles is constant its locus is a straight line parallel to the radical axis of the circles.

**20.** Prove that the polars of a fixed point with regard to a family of coaxal circles all pass through another fixed point.

**21.** The circles $\qquad x^2 + y^2 - 2ax \sec\alpha - a^2 = 0$
and $\qquad x^2 + y^2 - 2ay \operatorname{cosec}\alpha - a^2 = 0,$

where $\alpha$ is a given angle, both cut orthogonally every member of a coaxal family of circles. Find the radical axis and the limiting points of the family.

**22.** Prove that, if two points $P$, $Q$ are conjugate with regard to a circle, the circle on $PQ$ as diameter cuts the first circle orthogonally.

**23.** Prove that if $P$, $Q$ are conjugate points with regard to a circle, the circles with $P$, $Q$ as centres which cut the given circle orthogonally are orthogonal to one another.

**24.** Prove that, if $PQ$ is a diameter of a circle, then $P$, $Q$ are conjugate points with regard to any circle which cuts the given circle orthogonally.

**25.** Prove that, if $P$, $Q$ are conjugate points with regard to a circle, the square on $PQ$ is equal to the sum of the squares on the tangents from $P$, $Q$ to the circle.

**26.** The equation $x^2 + y^2 - 2g(x-1) = 5$, where $g$ is a variable parameter, represents a family of coaxial circles. Show that the radius of the smallest circle of the family is 2.

**27.** Prove that, if perpendiculars are drawn from a fixed point $P$ to the polars of $P$ with regard to a family of coaxial circles, then the locus of the feet of these perpendiculars is a circle whose centre lies on the radical axis of the family.

**28.** Prove that, if the points in which the line $lx + my + n = 0$ meets the circle $x^2 + y^2 + 2gx + 2fy + c = 0$, and those in which the line $l'x + m'y + n' = 0$ meets $x^2 + y^2 + 2g'x + 2f'y + c' = 0$ lie on a circle, then

$$2(g-g')(mn'-m'n) + 2(f-f')(nl'-n'l) + (c-c')(lm'-l'm) = 0.$$

**29.** Show that, if a diameter of a circle is the portion of the line $lx + my = 1$ intercepted by the lines $ax^2 + 2hxy + by^2 = 0$, then the equation of the circle is

$$(am^2 - 2hlm + bl^2)(x^2+y^2) + 2x(hm-bl) + 2y(hl-am) + a + b = 0.$$

**30.** Prove that, as $k$ varies, the equation

$$x^2 + y^2 + 2ax + 2by + c + 2k(ax - by + 1) = 0$$

represents a system of coaxial circles. Also prove that the orthogonal system is given by

$$x^2 + y^2 + \frac{c+2}{2a}\,x + \frac{c-2}{2b}\,y + h\left(\frac{x}{2a} + \frac{y}{2b} + 1\right) = 0,$$

where $h$ is a variable parameter.

# CHAPTER V

# THE PARABOLA

**5.1  Sections of a cone.** If a straight line $POP'$ revolves about a fixed axis $XOX'$ so that the lines are always inclined at the same angle, the line $POP'$ traces out a surface called a **right circular cone.** $O$ is the **vertex** and any position of the line $POP'$ is called a **generator** of the cone. It is assumed that the line $POP'$ can be produced indefinitely in either direction so that the conical surface is unlimited in extent in both directions. It can be seen that every point on $POP'$ describes a circle about a point on $XOX'$, so that every section of the cone by a plane at right angles to the axis is a circle; for example the section $A_1BA_1'B'$ in figure (i). Let $BB'$ be the diameter of this circle at right angles to the plane of the paper and suppose that the plane of the section is turned round the line $BB'$. The sections are at first closed curves called **ellipses**, as for example $A_2BA_2'B'$ in figure (ii), until the cutting plane becomes parallel to a generator $OP$; the section is then a curve called a **parabola**. Such a curve is shown in figure (iii), namely the curve $BA_3B'$. It is symmetrical about an axis through $A_3$ (its vertex) perpendicular to $BB'$ and it extends to infinity in the direction of its axis. As the plane of the section turns still further round $BB'$ it cuts both sheets of the cone as in figure (iv). This curve is called a **hyperbola** and has two distinct branches with an axis of symmetry $A_4A_4'$ and extends to infinity in both directions. If the cutting plane continues to turn round $BB'$ the points $A_4$, $A_4'$ move up to coincidence with the vertex $O$ of the cone and the section is then two straight lines $BO$, $B'O$, and further rotation of the plane will reproduce figures of the same types but in the reverse order—hyperbola, parabola, ellipse, circle.

These curves, ellipse, parabola and hyperbola, are called **conic sections** or conics and it can be demonstrated by pure geometry that a *conic section is the locus of a point which moves so that its* *distance from a fixed point bears a constant ratio to its distance from a fixed straight line. The fixed point is called the* **focus**, *the straight line the* **directrix**, *and the constant ratio the* **eccentricity**, *and the curve is an ellipse, parabola or hyperbola according as the eccentricity is less than, equal to or greater than unity.*

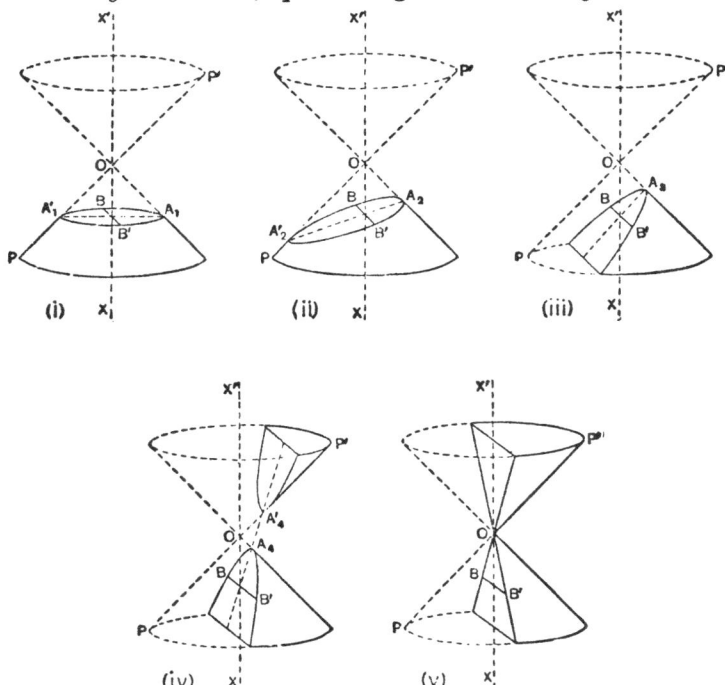

It can also be shown that sections of a cone by parallel planes are similar conics, i.e. conics having the same eccentricity. It follows that the forms of section mentioned above include all the possible forms. Also, for a reason that will appear later, a circle may be regarded as a special case of an ellipse. Planes of section which pass through the vertex of the cone give rise to degenerate forms. Such a plane may cut the cone in two distinct straight lines—a degenerate hyperbola; or, if a tangent plane, in two coincident straight lines—a degenerate parabola; or simply in a point—a degenerate ellipse. Consequently every section is either an ellipse, a parabola or a hyperbola.

**5.2   Equation of a parabola.** Let $S$ be the focus and $ZZ'$ the directrix and $P$ any point on the curve. Let $PM$ be perpendicular to $ZZ'$. The parabola is the locus of a point which moves so that $SP = PM$. Draw $SX$ perpendicular to $ZZ'$ and bisect it at $A$. Then $A$ is a point on the curve. Let $SA = AX = a$. Take $A$ as origin, the $x$-axis along $AS$, the $y$-axis parallel to the directrix and let $PN$ be the ordinate of $P$, $AN = x$, and $NP = y$.

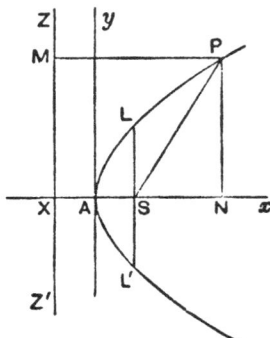

Then   $SP = MP = XN = a + x,$

and           $SN = x - a;$

but           $SP^2 = SN^2 + NP^2,$

therefore     $(a + x)^2 = (x - a)^2 + y^2,$

or            $y^2 = 4ax$           ......(1).

This is the equation of the parabola. The focus $S$ is the point $(a, 0)$, and the directrix $ZZ'$ is the line $x = -a$.

Since $y^2$ is positive it follows that, with a positive $a$, $x$ must be positive so that the curve lies wholly on one side of the axis of $y$. Also since for every positive value of $x$ there are two values of $y$, viz. $\pm 2\sqrt{(ax)}$, equal and opposite in sign, therefore the curve is symmetrical about the axis of $x$ and it extends to infinity as $x$ and $y$ increase together without limit. $A$ is called the **vertex** and the line $AS$ produced is called the **axis** of the curve. The double ordinate $LSL'$ through $S$ is called the **latus rectum**, its length is $4a$. It is the latus rectum or para-

meter in the equation $y^2 = 4ax$ that determines the size of a parabola.

**5.21   When** $a$ **is a** positive number, the equation           $y^2 = -4ax$

also represents a parabola, but it lies wholly on the negative side of the axis of $y$, the axis of the parabola coinciding with the negative direction of the $x$-axis.

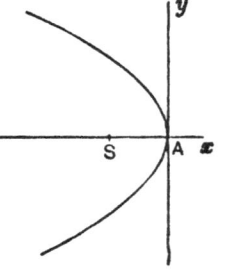

**5.211** If the focus $S$ of a parabola lies on the directrix $ZZ'$, say at $X$, then since every point $P$ for which $SP - PM$ lies on the perpendicular to the directrix through $S$, therefore the parabola is this perpendicular. This is the degenerate form of parabola referred to in 5.1 in which the plane of a parabolic section is moved parallel to itself until it passes through the vertex of the cone; it is then a tangent plane, i.e. meets the cone in two coincident straight lines which constitute the parabola. In this case the length of the latus rectum of the parabola is zero and its equation is $y^2 = 0$.

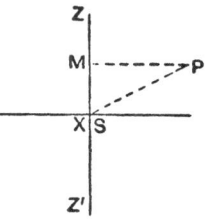

**5.22 Intersections of a straight line and a parabola.** The line $y = mx + c$ meets the curve $y^2 = 4ax$ in points whose abscissae are given by the equation
$$(mx + c)^2 = 4ax,$$

or
$$m^2x^2 + 2(mc - 2a)x + c^2 = 0 \qquad \ldots\ldots(1).$$

Since this is a quadratic in $x$ a straight line in general meets a parabola in two points. The points are real and distinct, coincident or imaginary according as
$$(mc - 2a)^2 > = \text{ or } < m^2c^2,$$

i.e. as
$$a > = \text{ or } < mc.$$

This line is therefore a tangent when $c = a/m$; i.e. the line
$$y = mx + \frac{a}{m} \qquad \ldots\ldots(2)$$

is a tangent to the parabola.

If we substitute $a/m$ for $c$ in (1), the equation becomes
$$m^2x^2 - 2ax + \frac{a^2}{m^2} = 0 \text{ or } \left(mx - \frac{a}{m}\right)^2 = 0,$$

showing that the point of contact of the tangent (2) is the point
$$x = \frac{a}{m^2}, \quad y = \frac{2a}{m}.$$

Returning to equation (1) we see that when $m$ is zero the line $y = c$ meets the curve in one point only at a finite distance, viz. at $\left(\frac{c^2}{4a}, c\right)$, the other root of the quadratic being infinite.

**5.3 Tangent to a parabola.** By differentiating with regard to $x$ the equation $y^2 = 4ax$, we get $y\frac{dy}{dx} = 2a$. Hence the gradient of the curve at a point $(x', y')$, or the value of $\dfrac{dy}{dx}$ at $(x', y')$, is

$\dfrac{2a}{y'}$. Therefore the equation of the tangent at $(x', y')$ is

$$y - y' = \frac{2a}{y'}(x - x'),$$

or
$$yy' = 2ax + y'^2 - 2ax',$$

and, since $(x', y')$ is on the curve, therefore

$$y'^2 = 4ax',$$

and the equation of the tangent reduces to

$$yy' = 2a(x + x') \qquad\qquad \ldots\ldots(1).$$

**5.31   Normal to a parabola.** Since the normal is at right angles to the tangent, therefore the equation of the normal at $(x', y')$ is $\qquad y'(x - x') + 2a(y - y') = 0.$

**5.4   Geometrical properties of a parabola.** Let $P$ be the point $(x', y')$ on the curve, and let the tangent and normal at $P$ cut the axis in $T$ and $G$.

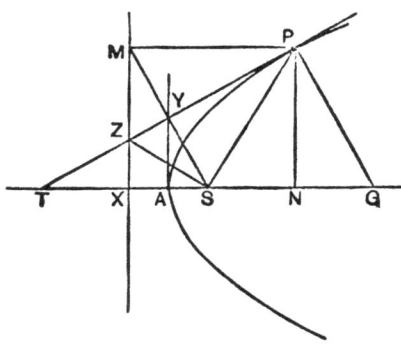

The equation of the tangent $TP$ is

$$yy' = 2a(x + x'),$$

and it meets the axis $y = 0$ where $x = -x'$, so that $AT = -AN$ or the subtangent $TN$ is bisected at the vertex.

Hence   $TS = TA + AS = AN + XA = XN = MP = SP.$

Therefore the angles $STP$, $SPT$ are equal. But $STP = MPT$ by parallels, therefore $MPT = SPT$, or the tangent bisects the angle between the focal distance $SP$ and the parallel to the axis $PM$.

Let *PT* meet the directrix in *Z* and *SM* in *Y*.

The triangles *SPZ*, *MPZ* can now be proved equal, so that the angle $PSZ = PMZ =$ a right angle.

The triangles *SPY*, *MPY* can also be proved equal so that *SM*, *PT* are at right angles and $SY = YM$.

But $SA = AX$, therefore *AY* is parallel to *XM* and therefore at right angles to the axis, so that *SM* meets the tangent at *P* at right angles on the tangent at the vertex.

Also the triangle *SYA* is similar to the triangle *STY* or *SPY*, so

that                     $SP:SY = SY:SA,$

or                       $SY^2 = SP.SA.$

Again the equation of the normal *PG* is

$$y'(x - x') + 2a(y - y') = 0,$$

and it meets the axis $y = 0$ in *G*, where

$$x = x' + 2a,$$

but $AN = x'$, therefore $NG = 2a$, i.e. the subnormal *NG* is constant and equal to half the latus rectum.

This last property also follows directly from the formula $y\frac{dy}{dx}$ (3.42) for the length of the subnormal at $(x, y)$.

**5.5  Parametric representation of the parabola.** For all values of *t* the point whose co-ordinates are

$$x = at^2, \quad y = 2at \qquad \ldots\ldots(1)$$

lies on the curve          $y^2 = 4ax,$

as can be seen at once by substituting the values of *x* and *y*.

The properties of a parabola can be investigated very simply by using the form (1) of the equation, which expresses *x* and *y* in terms of a variable parameter *t*.

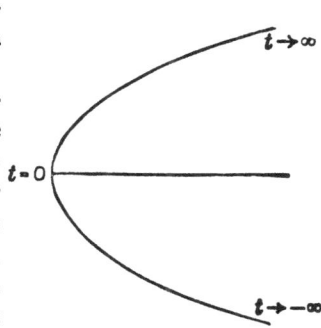

There is a point on the curve corresponding to every numerical value of *t* from $-\infty$ to $\infty$. The vertex is $t = 0$ and the lower and upper branches correspond to negative and positive values of *t*, and extend to $-\infty$ and $+\infty$; and in naming a point on the curve it is sufficient to call it the point $t$, it being understood that this means the point $(at^2, 2at)$.

**5.51 Chord joining the points $t_1$, $t_2$.** The line joining the points $(at_1{}^2, 2at_1)$ and $(at_2{}^2, 2at_2)$ has equation

$$\frac{x - at_1{}^2}{a\,(t_1{}^2 - t_2{}^2)} = \frac{y - 2at_1}{2a\,(t_1 - t_2)}.$$

Removing the factor $t_1 - t_2$, this reduces to

$$\tfrac{1}{2}y\,(t_1 + t_2) = x + at_1 t_2 \qquad \ldots\ldots(1).$$

**5.52 Tangent.** If the two points move up to coincidence, say at the point $t$, the chord becomes the tangent at $t$, and its equation is
$$yt = x + at^2,$$
or
$$y = \frac{x}{t} + at \qquad \ldots\ldots(2).$$

This equation might have been obtained directly, for by differentiation of
$$x = at^2, \quad y = 2at,$$
we get
$$dx = 2at\,dt \quad \text{and} \quad dy = 2a\,dt,$$
so that the gradient at $t$ is $\dfrac{dy}{dx} = \dfrac{1}{t}$, and the equation of the tangent is therefore
$$y - 2at = \frac{1}{t}\,(x - at^2) \quad \text{or} \quad y = \frac{x}{t} + at$$
as before.

We note the geometrical meaning of the parameter $t$. It is the reciprocal of the gradient of the curve, i.e. it is the cotangent of the angle which the tangent to the curve makes with the $x$-axis.

**5.53 The normal.** Since the slope of the tangent is $\dfrac{1}{t}$ that of the normal is $-t$ and its equation is
$$y - 2at = -t\,(x - at^2),$$
or
$$y + tx = 2at + at^3.$$

The parametric method is generally the most effective way of solving problems on the parabola and we shall now apply it to some examples.

**5.6 Examples.**

(i) *Find the co-ordinates of the point of intersection of the tangents at the points $t_1$, $t_2$ and prove that tangents at right angles intersect on the directrix.*

The tangents at $(at_1^2, 2at_1)$ and $(at_2^2, 2at_2)$ have equations

$$y = \frac{x}{t_1} + at_1 \text{ and } y = \frac{x}{t_2} + at_2.$$

Solving these equations for $x$ and $y$, we find that

$$x = at_1t_2 \text{ and } y = a(t_1 + t_2),$$

and these are the co-ordinates of the point of intersection of the tangents.

Further, the tangents are at right angles if the product of their $m$'s is $-1$, i.e. if $t_1t_2 = -1$. In this case for the point of intersection we have $x = -a$, so that the tangents intersect on the directrix.

(ii) *Three normals to a parabola can be drawn to pass through any point and the sum of the ordinates of the feet of the normals is zero.*

The equation of the normal at $(at^2, 2at)$ is

$$y + tx = 2at + at^3,$$

and this normal passes through a specified point $(h, k)$ if

$$k + th = 2at + at^3.$$

This equation is a cubic in $t$, viz.

$$at^3 + (2a - h)t - k = 0 \qquad \qquad \ldots\ldots(1),$$

and its roots $t_1$, $t_2$, $t_3$ give the points on the curve at which the normals pass through $(h, k)$.

Since the coefficient of $t^2$ in the equation is zero, therefore

$$t_1 + t_2 + t_3 = 0;$$

i.e. the sum of the ordinates $2a(t_1 + t_2 + t_3)$ is zero.

From equation (1) we also see that

$$t_2t_3 + t_2t_1 + t_1t_2 = 2 - \frac{h}{a}, \text{ and } t_1t_2t_3 = \frac{k}{a}.$$

These relations are useful in proving properties of the three normals from a point.

(iii) *Two fixed tangents to a parabola are cut proportionately by any variable tangent.*

Let $TP$, $TQ$ be the fixed tangents and let the tangent at $R$ cut them in $L$, $M$.

Let $t_1$, $t_2$ and $t$ be the parameters of $P$, $Q$, $R$.

Then using example (i) the abscissae of $T$, $L$ and $P$ are $at_1t_2$, $att_1$ and $at_1^2$ so that

$$TL:LP = att_1 - at_1t_2 : at_1^2 - att_1$$
$$= t - t_2 : t_1 - t.$$

Similarly the abscissae of $T$, $M$ and $Q$ are $at_1t_2$, $att_2$ and $at_2^2$ so that

$$TM:MQ = t - t_1 : t_2 - t.$$

Therefore

$$TL:LP = QM:MT.$$

### 5.61 Examples.

1. Show that the line $2y = 4x + a$ touches the parabola $y^2 = 4ax$, and find the co-ordinates of the point of contact.

2. Find the point of intersection of the parabolas $y^2 = 4ax$, $x^2 = 4ay$ other than the origin, and prove that the tangents at this point are inclined at an angle $\tan^{-1} \frac{3}{4}$.

3. Find the points in which the line $y = 8x - a$ cuts the parabola $y^2 = 4ax$, and find the point where the tangents at these points intersect.

4. A line through the vertex $A$ of a parabola $y^2 = 4ax$ makes an angle of $60°$ with the axis and cuts the curve again in $P$. Find the equation of the tangent at $P$, and show that the area of the triangle this tangent makes with the axes is $4a^2/3\sqrt{3}$.

5. Prove that in the figure of 5.4

$$SG = SP.$$

6. Prove that $t_1t_2 = -1$ is the condition that the chord joining the points $t_1$, $t_2$ on a parabola shall pass through the focus.

7. Prove that tangents drawn to a parabola from a point on the directrix are at right angles and that their chord of contact passes through the focus.

8. If in the figure of 5.4 $PS$ cuts the curve again in $Q$, prove that $QA$ passes through $M$.

9. Through the vertex $A$ of a parabola chords $AP$, $AQ$ at right angles to one another are drawn. Prove that $PQ$ cuts the axis in a fixed point.

**10.** Find the co-ordinates of the other point in which the normal at $(at^2, 2at)$ meets the parabola $y^2 = 4ax$; and prove that two normal chords that cut at right angles divide one another in the ratio $1:3$.

**11.** Three normals are drawn to a parabola from the point $(h, k)$. Prove that the centroid of the triangle formed by their feet is the point $(\frac{2}{3}(h - 2a), 0)$.

**12.** Find the equation of the tangent to the parabola $y^2 = 4ax$ that is parallel to the normal at $P$ $(at^2, 2at)$; and prove that, if this tangent meets the axis in $T$ and $PN$ is the ordinate of $P$ and $A$ is the vertex, then $$TA \cdot AN = a^2.$$

**13.** Prove that the circle $x^2 + y^2 + 2gx + 2fy + c = 0$ cuts the parabola $y^2 = 4ax$ in four points the sum of whose ordinates is zero; and conversely that if four points on a parabola be such that the sum of their ordinates is zero then the four points lie on a circle.

**14.** Prove that the orthocentre of a triangle whose sides all touch a parabola lies on the directrix.

**15.** A chord $POQ$ of a parabola $y^2 = 4ax$ cuts the axis in a fixed point $O$. $PN$, $QM$ are the ordinates of $P$ and $Q$, and $A$ is the vertex. Prove that $$NP \cdot MQ + 4a \cdot AO = 0.$$

**16.** From a point $P$ $(at_1^2, 2at_1)$ on the parabola $y^2 = 4ax$ two chords $PQ$, $PR$ are drawn, normal to the curve at $Q$ and $R$. Prove that, if $Q$, $R$ are the points $t_2$, $t_3$ on the curve, then $t_2t_3 = 2$, and the equation of $QR$ is $$yt_1 + 2(x + 2a) = 0.$$

**17.** Prove that the normals to a parabola at the ends of a chord whose inclination to the axis is $\theta$ meet on the normal whose inclination is $\tan^{-1}(2\cot\theta)$.

**18.** Prove that, if two parabolas are on the same side of the same directrix and have their axes in the same line, then they intersect at a distance from the directrix equal to one-quarter of the sum of their latera recta.

**5.7  Two tangents can be drawn from a point to a parabola.** The line $y = mx + \dfrac{a}{m}$ touches the parabola $y^2 = 4ax$, and it passes through a specified point $(h, k)$ if

$$k = mh + \frac{a}{m},$$

or if $$m^2h - mk + a = 0.$$

7

This quadratic in $m$ gives the directions of the tangents that pass through the point $(h, k)$ and they are real and distinct, coincident or imaginary according as $k^2 > =$ or $< 4ah$,

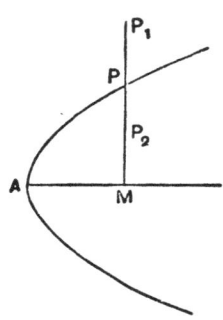

i.e. according as the point $(h, k)$ is outside, upon or inside the parabola.

For if the abscissa $AM = h$ then the ordinate $MP$ of a point on the curve is such that $MP^2 = 4ah$, and if $(h, k)$ is a point $P_1$ outside the parabola $k^2 = MP_1^2 > MP^2$, i.e. $k^2 > 4ah$, and if $(h, k)$ is a point $P_2$ inside the parabola $k^2 = MP_2^2 < MP^2$, i.e. $k^2 < 4ah$.

**5.71 Chord of contact of tangents from a point to a parabola.** The tangents to the parabola $y^2 = 4ax$ at the points $(h, k)$, $(h', k')$ have equations

$$yk = 2a (x + h)$$

and
$$yk' = 2a (x + h').$$

These tangents both pass through the point $(x', y')$ if

$$y'k = 2a (x' + h)$$

and
$$y'k' = 2a (x' + h').$$

These two equations show that the points $(h, k)$, $(h', k')$ lie on the line whose equation is

$$yy' = 2a (x + x') \qquad \ldots\ldots(1),$$

and this is therefore the equation of the chord of contact of tangents from $(x', y')$ to the parabola.

Equation (1) is the same in form as the equation of the tangent at $(x', y')$; so that the equation represents the tangent when $(x', y')$ is on the curve and the chord of contact of tangents when $(x', y')$ is not on the curve.

**5.72 To find the equation of the polar of the point $(x', y')$ with regard to the parabola $y^2 = 4ax$.** For definition of pole and polar see **4.51**.

Let $P$ be the point $(x', y')$ which may be inside or outside the parabola. Let *any* line through $P$ cut the parabola in $T$, $T'$ and let the tangents at $T$, $T'$ intersect in $Q$. Then the locus of $Q$ is the polar of $P$.

Let $x''$, $y''$ be the co-ordinates of $Q$, then by **5.71** the equation of $TT'$ is
$$yy'' = 2a\,(x + x''),$$
but $TT'$ passes through $P$ whose co-ordinates are $x'$, $y'$, therefore
$$y'y'' = 2a\,(x' + x'') \qquad \ldots\ldots(1).$$

Now writing current co-ordinates for $x''$, $y''$ we have for the locus of $Q$ the line
$$yy' = 2a\,(x + x') \qquad \ldots\ldots(2).$$

This is therefore the equation of the polar of $(x', y')$. The polar of a point with regard to a parabola is therefore a straight line and by comparing with **5.71** we see that when the point is outside the parabola its polar is the chord of contact of tangents from the point.

**5.73**  As in **4.54** it can be shown that *if the polar of P passes through Q, then the polar of Q passes through P, and that if the polars of P and Q intersect in R, then R is the pole of PQ.*

**5.74  Conjugate points and lines** are defined as in **4.55**. The points $(x', y')$, $(x'', y'')$ are conjugate with regard to the parabola $y^2 = 4ax$ if
$$y'y'' = 2a\,(x' + x'').$$

*To find the condition that the lines*
$$lx + my + n = 0, \quad l'x + m'y + n' = 0$$
*may be conjugate with regard to the parabola* $y^2 = 4ax$.

Let $(x', y')$ be the pole of $lx + my + n = 0$, then this equation, which represents the polar of $(x', y')$, must be the same as
$$yy' = 2a\,(x + x'),$$
so that by comparison   $\dfrac{-2a}{l} = \dfrac{y'}{m} = \dfrac{-2ax'}{n}$.

Hence the line $l'x + m'y + n' = 0$ passes through the pole $(x', y')$ of the line
$$lx + my + n = 0,$$
if
$$l'x' + m'y' + n' = 0,$$
or if
$$\frac{l'n}{l} - \frac{2amm'}{l} + n' = 0,$$
or if
$$ln' + l'n = 2amm',$$

which is the required condition.

**5.8 Locus of middle points of parallel chords.** Let $(x, y)$ be the co-ordinates of the middle point $V$ of a chord $QR$ which makes an angle $\theta$ with the $x$-axis. Let $Q$, $R$ be the points $(at_1^2, 2at_1)$, $(at_2^2, 2at_2)$. Then

$$\tan \theta = \frac{2a\,(t_1 - t_2)}{a\,(t_1^2 - t_2^2)} = \frac{2}{t_1 + t_2} \qquad \ldots\ldots(1).$$

And since $V$ is the middle point of $QR$, therefore

$$y = a\,(t_1 + t_2).$$

Hence, from (1), $\qquad y = 2a \cot \theta \qquad \ldots\ldots(2).$

If $\theta$ is constant, this equation represents a straight line parallel to the axis of the parabola, and it is the locus of the middle points of the parallel chords which make an angle $\theta$ with the axis.

*A line parallel to the axis of a parabola is called a* **diameter** *of the parabola, and the chords that it bisects are called ordinates to that diameter.*

**5.81 The tangent at the end of a diameter is parallel to the chords bisected by the diameter.** Let $PV$ be the diameter and $QVR$ a chord bisected by it. Let $QR$ move parallel to itself so that $V$ moves to $P$, then since the chord is always bisected by the diameter when $Q$ coincides with $P$ so does $R$ and the chord has then become the tangent at $P$.

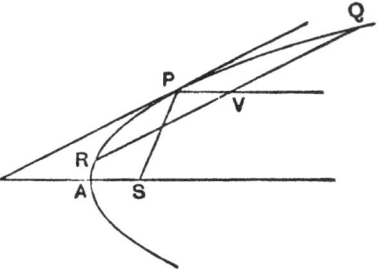

*Alternatively,* since at $P$ we have $y = 2a \cot \theta$, therefore the parameter $t$ of the point $P$ on the curve is $\cot \theta$, and the equation of the tangent at $P$ being

$$y = \frac{x}{t} + at$$

is $\qquad\qquad y = x \tan \theta + a \cot \theta,$

and makes the same angle $\theta$ with the $x$-axis as the parallel chords make.

**5.82 If the chord $QR$ is bisected at $V$ by the diameter $PV$ then $QV^2 = 4SP.PV$.** Let $Q$, $R$ be the points $(at_1^2, 2at_1)$, $(at_2^2, 2at_2)$. Then the co-ordinates of $V$ are $\frac{1}{2}a\,(t_1^2 + t_2^2)$, $a\,(t_1 + t_2)$, and by taking the squares of the differences of co-ordinates we get

$$QV^2 = \tfrac{1}{4}a^2\,(t_1^2 - t_2^2)^2 + a^2\,(t_1 - t_2)^2$$
$$= \tfrac{1}{4}a^2\,(t_1 - t_2)^2 \{(t_1 + t_2)^2 + 4\} \qquad \ldots\ldots(1).$$

Again if $P$ be the point $(at^2, 2at)$, since $PV$ is parallel to the axis

$$2at = a (t_1 + t_2), \quad \text{or} \quad t = \tfrac{1}{2} (t_1 + t_2);$$

therefore the co-ordinates of $P$ are $\tfrac{1}{4}a (t_1 + t_2)^2, a (t_1 + t_2)$.

Then, by subtracting $x$-co-ordinates of $P$ and $V$,

$$PV = \tfrac{1}{4}a (t_1 - t_2)^2 \qquad \ldots\ldots(2),$$

and since for any point $(x, y)$ we have $SP = x + a$, therefore in this case

$$SP = \tfrac{1}{4}a \{(t_1 + t_2)^2 + 4\} \qquad \ldots\ldots(3).$$

It follows from (1), (2) and (3) that

$$QV^2 = 4SP . PV.$$

**5.83  The tangents at the ends of a chord of a parabola intersect on the diameter that bisects the chord.**  With the notation of 5.82 the tangents at $Q$ and $R$ are

$$y = \frac{x}{t_1} + at_1 \quad \text{and} \quad y = \frac{x}{t_2} + at_2,$$

and these intersect, as we have seen in 5.6 (i), in the point

$$x = at_1t_2, \quad y = a (t_1 + t_2),$$

which from 5.8 lies on the diameter which bisects the chord.

Further, if $T$ be the point of intersection of the tangents we can prove that $TP = PV$. This follows at once from the $x$-co-ordinates of $T$, $P$ and $V$ found in this and the last articles.

**5.9  Examples.**

(i) *Find the equation of the chord of the parabola $y^2 = 4ax$ which has $(x', y')$ as middle point.*

The equation of the chord joining the points $t_1$, $t_2$ is

$$\tfrac{1}{2}y (t_1 + t_2) = x + at_1t_2 \quad (5.51).$$

The co-ordinates of the middle point of the chord are

$$\tfrac{1}{2}a (t_1^2 + t_2^2) = x', \quad \text{and} \quad a (t_1 + t_2) = y',$$

whence                $2a^2 t_1 t_2 = y'^2 - 2ax'$.

Hence, by eliminating $t_1$ and $t_2$ we find for the equation of the chord

$$yy' - 2ax = y'^2 - 2ax'.$$

(ii) *The polar of the point $P$ with regard to the parabola $y^2 = 4ax$ meets the curve in $Q$, $R$. Show that if $P$ lies on the line*

$$lx + my + n = 0,$$

*then the middle point of $QR$ lies on the parabola*

$$l (y^2 - 4ax) + 2a (lx + my + n) = 0.$$

Let $P$ be the point $(x'', y'')$ and let $(x', y')$ be the middle point of $QR$. Then as in the last example the equation of $QR$ is

$$yy' - 2ax = y'^2 - 2ax';$$

but this is the polar of $P$ and is therefore the same equation as

$$yy'' = 2a (x + x'').$$

Whence by comparing the equations we get

$$y'' = y' \text{ and } x'' = -x' + \frac{y'^2}{2a}.$$

But $(x'', y'')$ lies on $lx + my + n = 0$, whence by substitution we get

$$l\left(-x' + \frac{y'^2}{2a}\right) + my' + n = 0,$$

so that the locus of $(x', y')$ is

$$l(y^2 - 4ax) + 2a(lx + my + n) = 0.$$

**5.91 Examples.** In the following examples the equation of a parabola is assumed to be $y^2 = 4ax$ and capital letters refer to the figure of 5.4 unless the contrary is stated.

1. Prove that as $P$ moves along the curve $GP^2$ varies as $SG$.

2. Prove that, if $PP'$ is a double ordinate and $PX$ meets the curve in $Q$, then $P'Q$ passes through $S$.

3. Prove that, if $PSP'$ is a focal chord and $AP$, $AP'$ meet the latus rectum in $Q$, $Q'$, then $SQ$, $SQ'$ are equal to the ordinates of $P'$ and $P$.

4. Prove that, if the tangents at $P$, $Q$ intersect in $T$, then

$$ST^2 = SP.SQ.$$

5. Prove that, if the tangent at the end $Q$ of a focal chord $PSQ$ meets the latus rectum in $R$, then $PGR$ is a right angle.

6. Tangents at $P$, $Q$, $R$ on a parabola form a triangle $UVW$. Show that the centroids of the triangles $PQR$ and $UVW$ lie on the same diameter.

7. Prove that, if the difference of the ordinates of two points on a parabola is constant, then the locus of the point of intersection of the tangents at these points is an equal parabola.

8. Prove that, if two tangents intercept a fixed length on the tangent at the vertex, the locus of their intersections is an equal parabola.

**9.** The chord of contact of tangents from any point $Q$ meets the tangent at the vertex in $R$. Prove that the tangent of the angle which $AQ$ makes with the axis is $2a/AR$.

**10.** The parameters $t$, $t'$ of two points on a parabola are connected by the relation $t = k^2t'$, prove that the tangents at the points intersect on the curve

$$y^2 = \left(k + \frac{1}{k}\right)^2 ax.$$

**11.** Show that the length of the normal chord at the point of parameter $t$ is $\qquad 4a\,(1 + t^2)^{\frac{3}{2}}/t^2$.

**12.** Prove that the locus of intersection of tangents at the ends of a normal chord is $\qquad (x + 2a)\,y^2 + 4a^3 = 0$.

**13.** Prove that the locus of the point of intersection of perpendicular normals is the parabola $y^2 = a\,(x - 3a)$.

**14.** Prove that if the tangents at two points on the parabola intersect in the point $(x', y')$ the corresponding normals intersect in the point

$$\left(2a - x' + \frac{y'^2}{a},\ -\frac{x'y'}{a}\right).$$

**15.** Show that, if the tangent at $P$ meets the latus rectum in $K$, then $SK$ is a mean proportional between the segments of the focal chord through $P$.

**16.** Show that, if the tangents from $Q$ to the parabola form with the tangent at the vertex a triangle of constant area $c^2$, then the locus of $Q$ is the curve $\qquad x^2\,(y^2 - 4ax) = 4c^4$.

**17.** Show that normals at the ends of each of a series of parallel chords of a parabola intersect on a fixed straight line, itself a normal to the parabola.

**18.** $P$, $Q$ are points on the parabola subtending a constant angle $\alpha$ at the vertex. Show that the locus of the intersection of the tangents at $P$, $Q$ is the curve $\quad 4\,(y^2 - 4ax) = (x + 4a)^2 \tan^2 \alpha$.

**19.** Prove that the exterior angle between two tangents to a parabola is equal to the angle which either of them subtends at the focus.

**20.** Two perpendicular focal chords of a parabola meet the directrix in $T$ and $T'$; show that the tangents to the parabola which are parallel to these chords intersect in the middle point of $TT'$.

**21.** Prove that, if the tangents at the points $Q$, $R$ intersect at $P$, then
$$PQ^2 : PR^2 = SQ : SR.$$

**22.** The tangents at any two points $P$, $Q$ meet at $T$ and the normals meet at $N$. Prove that the projection of $TN$ on the axis is equal to the sum of the distances of $P$ and $Q$ from the directrix.

**23.** Prove that the circumscribing circle of the triangle formed by three tangents to a parabola passes through the focus.

**24.** $PQ$ is a chord of a parabola normal at $P$; the circle on $PQ$ as diameter cuts the parabola again in $R$. Prove that the projection of $QR$ on the axis is twice the latus rectum.

**25.** Prove that the distance between a tangent and the parallel normal is $a \operatorname{cosec} \theta \sec^2 \theta$, where $\theta$ is the angle which either makes with the axis.

**26.** Prove that, if the normals at $P$ and $Q$ intersect on the curve, then $PQ$ cuts the axis in a fixed point.

**27.** Prove that, if the normals at $P$ and $Q$ meet at the point $R$ $(x', y')$ on the parabola, and the tangents at $P$ and $Q$ meet at $T$, then
$$TP \cdot TQ = \tfrac{1}{2}(x' - 8a) \sqrt{(y'^2 + 4a^2)}.$$

**28.** Show that, in the last example, as $R$ moves along the parabola, the middle point of $PQ$ always lies on the parabola
$$y^2 = 2a(x + 2a).$$

**29.** Prove that the area of the triangle formed by the tangents at the points $t_1$, $t_2$ and their chord of contact is
$$\tfrac{1}{2}a^2(t_1 - t_2)^3.$$

**30.** Prove that the area of the triangle formed by three points $t_1$, $t_2$, $t_3$ on the parabola is
$$a^2(t_2 - t_3)(t_3 - t_1)(t_1 - t_2);$$
and that this is double the area of the triangle formed by the tangents at these points.

**31.** Prove that, if a line through any point $P$ $(x', y')$ making an angle $\theta$ with the axis meets the parabola at $Q$ and $R$, then
$$PQ \cdot PR = (y'^2 - 4ax') \operatorname{cosec}^2 \theta.$$

**32.** Two chords $QR$, $Q'R'$ of a parabola meet at $O$, and the diameters bisecting them meet the curve at $P$ and $P'$. Prove that
$$QO \cdot OR : Q'O \cdot OR' = SP : SP'.$$

**33.** Show that, if $P$ is on the parabola, the length of the chord through $P$ that makes an angle $\theta$ with the axis is

$$4a \sin (\alpha - \theta) \operatorname{cosec}^2 \theta \operatorname{cosec} \alpha,$$

where $\alpha$ is the inclination of the tangent at $P$ to the axis.

**34.** Show that the locus of the middle point of a chord which passes through the fixed point $(h, k)$ is the parabola

$$y^2 - ky = 2a (x - h).$$

**35.** A tangent to the parabola $y^2 + 4bx = 0$ meets the parabola $y^2 = 4ax$ at $P, Q$. Prove that the locus of the middle point of $PQ$ is

$$y^2 (2a + b) = 4a^2 x.$$

**36.** Prove that the polar of the focus of a parabola is the directrix.

**37.** Prove that, if a chord of the parabola subtends a right angle at the vertex, the locus of its pole is $x + 4a = 0$.

**38.** Show that, if parabolas $y^2 = 4ax$ are drawn corresponding to different values of $a$, the feet of the perpendiculars from a fixed point on its polar lines all lie on a circle passing through the point.

**39.** Prove that, if from a point $Q (x', y')$ a perpendicular be drawn to the polar of $Q$ with regard to the parabola cutting it in $R$ and the axis in $G$, then $SG = SR = x' + a$.

**40.** Prove that, if the diameter through a point $P$ of a parabola meets any chord in $O$ and the tangents at the ends of the chord in $T, T'$, then $PO^2 = PT \cdot PT'$.

**41.** $QQ'$ is a chord of a parabola and $TOR$ is a diameter which meets the tangent at $Q$ in $T$, the curve in $O$ and $QQ'$ in $R$. Prove that

$$TO : OR = QR : RQ'.$$

**42.** Prove that if the normals at three points $P, Q, R$ on a parabola concur, then the points $P, Q, R$ and the vertex of the parabola are concyclic.

**43.** Prove that, in general, two members of the family of parabolas $y^2 = 4a (x + a)$, where $a$ is the parameter specifying members of the family, pass through any assigned point of the plane, and that these two parabolas cut orthogonally at $P$.

# THE ELLIPSE

**6.1 DEF.** An ellipse is the locus of a point which moves so that the ratio of its distance from a fixed point to its distance from a fixed straight line is a constant, less than unity.

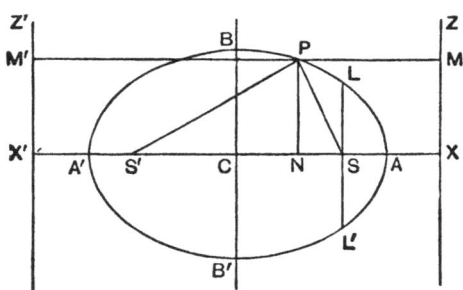

The fixed point is called the *focus*, the fixed line the *directrix* and the constant ratio the *eccentricity*. The latter is denoted by *e*.

*To find the equation of an ellipse.*

Let $S$ be the focus and $ZX$ the directrix. Draw $SX$ perpendicular to the directrix and divide it internally and externally at $A$ and $A'$ in the ratio $e:1$, so that $A$ and $A'$ are points on the locus.

Let $C$ be the middle point of $A'A$ and let $A'A = 2a$. Then we have

$$e = \frac{A'S}{A'X} = \frac{SA}{AX} = \frac{A'S+SA}{A'X+AX} = \frac{A'S-SA}{A'X-AX} = \frac{2CA}{2CX} = \frac{2CS}{2CA}.$$

Therefore $CX = CA/e = a/e$ and $CS = eCA = ea$ ......(1).

Take $C$ as origin and the $x$-axis along $CX$, and let $x$, $y$ be the co-ordinates of any point $P$ on the locus, $NP$ the ordinate of $P$, and $PM$ the perpendicular to the directrix.

Then $$SP^2 = e^2PM^2 = e^2NX^2,$$

therefore $$NS^2 + NP^2 = e^2(CX - CN)^2;$$

but $$NS = CS - CN = ae - x,$$

therefore $\qquad (ae - x)^2 + y^2 = e^2 \left( \dfrac{a}{e} - x \right)^2 = (a - ex)^2,$

or $\qquad x^2 (1 - e^2) + y^2 = a^2 (1 - e^2),$

and by dividing by $a^2 (1 - e^2)$ this gives

$$\frac{x^2}{a^2} + \frac{y^2}{a^2 (1 - e^2)} = 1 \qquad \ldots\ldots(2).$$

The intercepts on the axis of $y$ found by putting $x = 0$ are $y = \pm a \sqrt{(1 - e^2)}$, and denoting these lengths by $\pm b$, we have

$$b^2 = a^2 (1 - e^2) \qquad \ldots\ldots(3),$$

and equation (2) may be written

$$\frac{x^2}{a^2} + \frac{y^2}{b^2} = 1 \qquad \ldots\ldots(4),$$

which is the standard form of equation of an ellipse.

The intercepts on the axes, viz. $A'A$ and $B'B$, are called the **major and minor axes** of the ellipse. They are of lengths $2a$ and $2b$ respectively. The double ordinate $LSL'$ through the focus is called the **latus rectum**.

Since $\quad SL = eSX = e (CX - CS) = a (1 - e^2) = b^2/a \ \ldots\ldots(5),$
therefore $b^2/a$ or $a (1 - e^2)$ is the length of the semi-latus rectum.

From (4) we have $y = \pm \dfrac{b}{a} \sqrt{(a^2 - x^2)}$, so that for real values of $y$ we cannot have $x$ greater than $a$, similarly $y$ cannot be greater than $b$, so that the curve is limited in the directions of both axes.

**6.11 The second focus and directrix.** Since the equation of the ellipse is unaltered by changing the sign of $x$ or of $y$ it follows that the curve is symmetrical with regard to both axes; i.e. if $(x, y)$ is a point on the curve then the points $(x, -y)$, $(-x, y)$ and $(-x, -y)$ are also on the curve. And since the line joining $(x, y)$ and $(-x, -y)$ goes through the origin, it follows that every chord through $C$ is bisected at $C$. $C$ is called the **centre** of the ellipse.

Again, from the symmetry it follows that if on $CA'$ we take points $S'$ and $X'$ such that $CS' = eCA'$ and $CX' = CA'/e$ and describe an ellipse with $S'$ as focus and a directrix $Z'X'$ parallel to $ZX$ and $S'P = ePM'$ it will be the same ellipse as that obtained with $S$ as focus and $ZX$ as directrix, $S'$ and $Z'X'$ are therefore the second focus and directrix.

*SP and S'P are called the focal distances and their sum is equal to the major axis.*

For $\qquad SP = ePM = eNX = e\,(CX - CN) = a - ex$

and $\qquad S'P = eM'P = eX'N = e\,(X'C + CN) = a + ex,$

so that $\qquad\qquad SP + S'P = 2a.$

This suggests a mechanical method of drawing ellipses. If two drawing pins are stuck into the paper at $S$ and $S'$ and a loop of thread is placed round them and kept taut by the point of a pencil at $P$, so that the thread forms the sides of the triangle $PS'S$, then the point of the pencil can be moved keeping the thread taut only in such a way that $SP + PS'$ remains constant; i.e. $P$ will describe an ellipse of foci $S$, $S'$ and major axis $SP + PS'$.

**6.12** By considering points $P_1$, $P$, $P_2$ on the same ordinate and outside, upon and inside the ellipse respectively, it is easy to see as in **5.7** that the point $(x, y)$ is outside, upon or inside the ellipse according as $\frac{x^2}{a^2} + \frac{y^2}{b^2} - 1$ is positive, zero or negative.

**6.13 The circle is a limiting form of ellipse.** If the eccentricity of an ellipse is diminished to zero while the major axis remains constant, then since $CS = eCA$ the foci move up to the centre and $b$ becomes equal to $a$ and the figure becomes a circle of radius $a$.

**6.14 The parabola is a limiting form of ellipse.** In the figure of **6.1** let the origin be moved to $A'$, i.e. to the point $(-a, 0)$. The equation of the ellipse becomes

$$\frac{(x - a)^2}{a^2} + \frac{y^2}{b^2} = 1 \quad (3.6),$$

or $\qquad\dfrac{x^2}{a^2} - \dfrac{2x}{a} + \dfrac{y^2}{a^2(1 - e^2)} = 0.$

Now keep $A'S'$ or $a\,(1 - e)$ fixed and $= c$ say, and let the eccentricity $e$ increase to unity. This means that as $1 - e$ tends to zero $a$ must increase and tend to infinity.

But $\qquad y^2 = 2ax\,(1 - e^2) - x^2\,(1 - e^2)$

$$= 2cx\,(1 + e) - \frac{cx^2}{a}\,(1 + e).$$

Now as $e \to 1$ and $a \to \infty$ this takes the form

$$y^2 = 4cx,$$

which is a parabola of latus rectum 4 times $S'A'$.

**6.15** If the focus $S$ of an ellipse lies on the directrix $ZX$, then since in a right-angled triangle $SPM$, the hypotenuse $SP$ is greater than $PM$, therefore the relation $\dfrac{SP}{PM} = e < 1$, can only be satisfied by $SP = PM = 0$, i.e. the locus degenerates into the point $S$. This is the case when the plane of section of a cone passes through the vertex and meets the cone in this point only, i.e. in a "point ellipse" (see **5.1**).

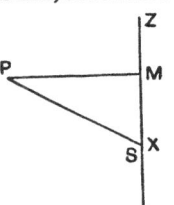

**6.2 Intersections of a straight line and an ellipse.** The line $y = mx + c$ meets the ellipse $\dfrac{x^2}{a^2} + \dfrac{y^2}{b^2} = 1$, in points whose abscissae are given by the equation

$$\frac{x^2}{a^2} + \frac{(mx + c)^2}{b^2} = 1,$$

or $\qquad x^2 (a^2m^2 + b^2) + 2a^2cmx + a^2 (c^2 - b^2) = 0 \qquad \ldots\ldots(1).$

This being a quadratic in $x$ it follows that the straight line meets the ellipse in two points which are real and distinct, coincident or imaginary according as

$$a^4c^2m^2 - a^2 (a^2m^2 + b^2)(c^2 - b^2) > = \quad \text{or} \quad < 0,$$

i.e. according as $\quad a^2m^2 + b^2 > = \quad \text{or} \quad < c^2$.

The roots are therefore equal and the line touches the ellipse when $\qquad c = \pm \sqrt{(a^2m^2 + b^2)},$

so that the lines $\qquad y = mx \pm \sqrt{(a^2m^2 + b^2)} \qquad \ldots\ldots(2)$

are tangents to the ellipse.

If we rationalise the last equation we may write it as a quadratic in $m$, viz. $\qquad m^2 (x^2 - a^2) - 2mxy + y^2 - b^2 = 0 \qquad \ldots\ldots(3);$

and if we now regard $(x, y)$ as a fixed point this quadratic gives the directions of the tangents that can be drawn to the ellipse from $(x, y)$. Since the equation in $m$ is a quadratic it follows that two tangents can be drawn from any point to an ellipse. The tangents are real and

distinct, coincident or imaginary along with the roots of the quadratic,
i.e. according as $\qquad x^2y^2 - (x^2 - a^2)(y^2 - b^2)$

is positive, zero or negative or as

$$\frac{x^2}{a^2} + \frac{y^2}{b^2} - 1$$

is positive, zero or negative; i.e. according as the point $(x, y)$ is outside,
upon or inside the ellipse.

**6.21**  Director circle.  *The locus of the point of intersection of two
tangents at right angles is a circle.*

From **6.2** (3) the directions of the tangents that pass through a
point $(x, y)$ are given by the quadratic equation

$$m^2(x^2 - a^2) - 2mxy + y^2 - b^2 = 0.$$

If $m_1$, $m_2$ are the roots of this quadratic, they are the $m$'s of lines at
right angles to one another if $m_1m_2 = -1$, or if

$$\frac{y^2 - b^2}{x^2 - a^2} = -1, \text{ or } x^2 + y^2 = a^2 + b^2.$$

Hence the circle $\qquad x^2 + y^2 = a^2 + b^2$

is the locus of the point of intersection of tangents to the ellipse which
are at right angles to one another.  It is called the *Director Circle*.

**6.22**  **Tangent to an ellipse at the point** $(x', y')$.  By
differentiating the equation $\dfrac{x^2}{a^2} + \dfrac{y^2}{b^2} = 1$ we get

$$\frac{x}{a^2} + \frac{y}{b^2}\frac{dy}{dx} = 0.$$

Therefore the gradient of the curve at $(x', y')$ is $-\dfrac{b^2x'}{a^2y'}$, and
the equation of the tangent at that point is

$$y - y' = -\frac{b^2x'}{a^2y'}(x - x'),$$

or $\qquad\qquad \dfrac{xx'}{a^2} + \dfrac{yy'}{b^2} = \dfrac{x'^2}{a^2} + \dfrac{y'^2}{b^2};$

but $\dfrac{x'^2}{a^2} + \dfrac{y'^2}{b^2} = 1$, since $(x', y')$ is on the ellipse, therefore the
equation of the tangent is

$$\frac{xx'}{a^2} + \frac{yy'}{b^2} = 1 \qquad\qquad \ldots\ldots(1).$$

Another method of finding the equation of the tangent is given in **6.41**.

**6.23    Condition that the line $lx + my + n = 0$ may touch the ellipse.** If the line is the tangent at $(x', y')$ the equation must be the same as $\dfrac{xx'}{a^2} + \dfrac{yy'}{b^2} = 1$, so that by comparing coefficients we must have

$$\frac{x'}{a^2 l} = \frac{y'}{b^2 m} = \frac{-1}{n}.$$

But
$$\frac{x'^2}{a^2} + \frac{y'^2}{b^2} = 1,$$

therefore    $$a^2 l^2 + b^2 m^2 = n^2 \qquad \ldots\ldots(1)$$

is the required condition.

In the same way it may be shown that the line

$$x \cos \alpha + y \sin \alpha = p$$

is a tangent if    $$a^2 \cos^2 \alpha + b^2 \sin^2 \alpha = p^2 \qquad \ldots\ldots(2).$$

**6.231    Perpendicular distance $p$ of tangent from centre.** Equation (2) of **6.23** gives the central perpendicular on the tangent in terms of the angle the perpendicular makes with the major axis.

In terms of $(x', y')$ the perpendicular from the origin to the line

$$\frac{xx'}{a^2} + \frac{yy'}{b^2} = 1$$

is
$$p = \frac{1}{\sqrt{\left(\dfrac{x'^2}{a^4} + \dfrac{y'^2}{b^4}\right)}},$$

or in a more convenient form

$$\frac{1}{p^2} = \frac{x'^2}{a^4} + \frac{y'^2}{b^4}.$$

**6.24    Equation of the normal at $(x', y')$.** Since the normal is perpendicular to the tangent whose equation is (1) **6.22**, its equation can be written in the symmetrical form

$$\frac{x - x'}{\dfrac{x'}{a^2}} = \frac{y - y'}{\dfrac{y'}{b^2}} \qquad \ldots\ldots(1).$$

In working problems on the normal it is sometimes useful to carry the equation further and, using an algebraical theorem about equal ratios, write

$$\frac{x-x'}{\dfrac{x'}{a^2}}=\frac{y-y'}{\dfrac{y'}{b^2}}=\pm\frac{\sqrt{\{(x-x')^2+(y-y')^2\}}}{\sqrt{\left(\dfrac{x'^2}{a^4}+\dfrac{y'^2}{b^4}\right)}}.$$

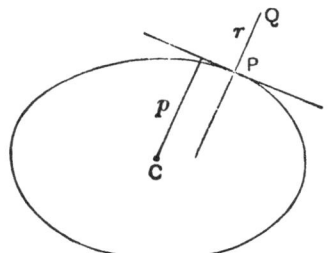

Then if $Q$ be the point $(x, y)$ on the normal and $r$ its distance from $P$ the numerator of the last fraction is $r$, and the denominator is $1/p$ (6.231), so that for any point $(x, y)$ on the normal at $(x', y')$ we have

$$\frac{x-x'}{\dfrac{x'}{a^2}}=\frac{y-y'}{\dfrac{y'}{b^2}}=\pm\, rp \qquad\qquad\ldots\ldots(2),$$

and (by considering points in the positive quadrant) the positive sign must be taken if $x - x'$ and $y - y'$ are positive, i.e. when $(x, y)$ is on the outward drawn normal, and the negative sign when $(x, y)$ is on the inward drawn normal.

**6.25 Examples.** When the focus, corresponding directrix and eccentricity of an ellipse are given the equation may be written down directly from the definition: e.g.

(i) *Find the equation of an ellipse of eccentricity $\frac{3}{4}$ whose focus is the point $(2, -1)$ and corresponding directrix the line $x - y = 0$. Find also the lengths of the axes.*

If $(x, y)$ is any point on the curve its distance from the point $(2, -1)$ is $\frac{3}{4}$ of its distance from the line $x - y = 0$.

Therefore $$(x-2)^2+(y+1)^2=\frac{9}{16}\cdot\frac{(x-y)^2}{2}$$

is the equation. It reduces to

$$23x^2+18xy+23y^2-128x+64y+160=0.$$

Again, if $S$ be the focus and $SX$ the perpendicular on the directrix $x - y = 0$,

$$SX = \frac{2+1}{\sqrt{2}} = \frac{3}{\sqrt{2}};$$

also $$\frac{SA}{3} = \frac{AX}{4} = \frac{SX}{7}, \text{ and } \frac{A'S}{3} = \frac{A'X}{4} = \frac{SX}{1}.$$

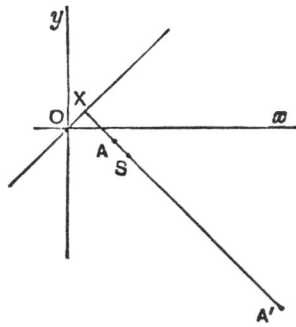

Therefore $$AA' = AS + SA' = \frac{24}{7} \cdot SX = \frac{72}{7\sqrt{2}}.$$

Also $$b^2 = a^2 (1 - e^2) = \left(\frac{36}{7\sqrt{2}}\right)^2 \left(1 - \frac{9}{16}\right),$$

so that the minor axis $$2b = 18/\sqrt{14}.$$

(ii) *Find the equation of an ellipse of eccentricity $\frac{1}{2}$, which has its major and minor axes along $Ox$ and $Oy$ and touches the line $x + 2y = 5$.*

Let the equation be $\dfrac{x^2}{a^2} + \dfrac{y^2}{b^2} = 1$. Then $b^2 = a^2 (1 - e^2) = \frac{3}{4}a^2$.

And taking the condition of tangency of the line $lx + my + n = 0$ to be $a^2l^2 + b^2m^2 = n^2$ (6.23 (1)), in this case it is

$$a^2 + 4b^2 = 25, \text{ or } 4a^2 = 25;$$

therefore $a^2 = \frac{25}{4}$ and $b^2 = \frac{75}{16}$, and the required ellipse is

$$12x^2 + 16y^2 = 75.$$

**6.26 Examples.**

**1.** Find the length of the latus rectum, the eccentricity and the co-ordinates of the foci of the ellipses

(i) $4x^2 + 9y^2 = 36$; (ii) $9x^2 + 4y^2 = 36$; (iii) $x^2 + 2y^2 = 8$.

**2.** Find the equation of the ellipse whose foci are the points $(3, 0)$, $(-3, 0)$ and eccentricity $\frac{1}{2}$. What are the equations of the directrices?

**3.** Find the equation of the ellipse of eccentricity $\frac{1}{4}$, which has its centre at the point $(4, 0)$ and touches the axis of $y$ at the origin. What is the length of its latus rectum?

**4.** An ellipse has the axis of $y$ for directrix and its centre at the point $(6, 0)$. Find its equation if its eccentricity is $\frac{2}{3}$.

**5.** Find the equation of the ellipse which has a focus at $(3, 1)$, corresponding directrix the line $x = 6$ and eccentricity $\frac{1}{2}$. What are the lengths of its axes?

**6.** An ellipse has its centre at $(2, 3)$ and a focus at $(3, 4)$ and its eccentricity is $\frac{1}{2}$. Find its equation.

**7.** Are the points $(2, 1)$, $(3, -2)$ inside or outside the ellipse
$$2x^2 + 5y^2 - 20 = 0?$$

**8.** Prove that the line $y = x - 5$ touches the ellipse
$$9x^2 + 16y^2 = 144,$$
and find the co-ordinates of the point of contact.

**9.** The line $2x + 3y = c$ touches the ellipse $4x^2 + 9y^2 = 1$. Find the values of $c$.

**10.** Find the equation of the tangent to the ellipse $x^2 + 4y^2 = 8$ at the point $(2, 1)$, and also the equations of the two tangents perpendicular to this.

**11.** Find the equations of the tangents to the ellipse $3x^2 + 4y^2 = 2$ which makes an angle of $60°$ with the major axis.

**12.** Find at what point the line $y + ex = a$ touches the ellipse
$$x^2 (1 - e^2) + y^2 = a^2 (1 - e^2).$$

**13.** Prove that the normal at an end of the latus rectum meets the major axis at a distance $ae^3$ from the centre.

**14.** Prove that, if the normal at an end $L$ of the latus rectum meets the minor axis in $g$, and $l$ is the projection of $L$ on the minor axis, then $gl = a$.

**15.** Prove that the normal at $(x, y)$ divides the major axis into segments of lengths $a - e^2x$ and $a + e^2x$.

**16.** Prove that, if the normal at $P$ meets the major axis in $G$, and the minor axis in $G'$, then

(i) $SG = eSP$;

(ii) $PG : PG' = b^2 : a^2$.

**6.3 Chord of contact of tangents.** To find the equation of the chord of contact of the tangents drawn from the point $(x', y')$. Let $(h, k)$ and $(h', k')$ be the points of contact of the tangents, then their equations are

$$\frac{xh}{a^2} + \frac{yk}{b^2} = 1 \text{ and } \frac{xh'}{a^2} + \frac{yk'}{b^2} = 1 \quad (6.22);$$

and since these lines both pass through the point $(x', y')$ therefore

$$\frac{x'h}{a^2} + \frac{y'k}{b^2} = 1 \text{ and } \frac{x'h'}{a^2} + \frac{y'k'}{b^2} = 1.$$

The last two equations show that $(h, k)$ and $(h', k')$ lie on the line whose equation is

$$\frac{xx'}{a^2} + \frac{yy'}{b^2} = 1 \qquad \qquad \ldots\ldots(1).$$

This is therefore the equation of the chord of contact of the tangents from $(x', y')$. The equation (1) is the same in form as the equation of the tangent at $(x', y')$; so that the equation represents the tangent when the point $(x', y')$ is on the curve and the chord of contact of tangents when $(x', y')$ is not on the curve.

**6.31 Pole and polar.** For definition of pole and polar see **4.51**.

As in **4.51** or **5.72** it can be shown that the polar of the point $(x', y')$ is the straight line $\frac{xx'}{a^2} + \frac{yy'}{b^2} = 1$.

As in **4.54** it can be shown that *if the polar of P passes through Q, then the polar of Q passes through P, and that if the polars of P and Q intersect in R, then R is the pole of PQ.*

**6.32 Conjugate points and lines** are defined as in **4.55**. The points $(x', y')$ and $(x'', y'')$ are conjugate with regard to the ellipse

$$\frac{x^2}{a^2} + \frac{y^2}{b^2} = 1, \text{ if } \frac{x'x''}{a^2} + \frac{y'y''}{b^2} = 1$$

*To find the condition that the lines*

$$lx + my + n = 0, \quad l'x + m'y + n' = 0$$

*may be conjugate with regard to the ellipse* $\frac{x^2}{a^2} + \frac{y^2}{b^2} = 1$.

Let $(x', y')$ be the pole of

$$lx + my + n = 0,$$

then this equation which represents the polar of $(x', y')$ must be the same as

$$\frac{xx'}{a^2} + \frac{yy'}{b^2} = 1,$$

so that by comparison $\quad \frac{x'}{a^2 l} = \frac{y'}{b^2 m} = -\frac{1}{n}.$

Hence the line $l'x + m'y + n' = 0$ passes through the pole $(x', y')$ of the line

$$lx + my + n = 0,$$

if $\qquad\qquad l'x' + m'y' + n' = 0,$

or if $\qquad -\dfrac{a^2 ll'}{n} - \dfrac{b^2 mm'}{n} + n' = 0,$

or if $\qquad a^2 ll' + b^2 mm' = nn',$

which is the required condition.

**6.33** We have now seen how to find the equation of the tangent and the polar of a point with regard to a circle, a parabola and an ellipse and we observe that for any of these curves given by an equation of the second degree the equation of the tangent at $(x', y')$ or of the polar of $(x', y')$ may be obtained directly from the equation of the curve by changing $x^2$ into $xx'$, $y^2$ into $yy'$, $2x$ into $x + x'$, and $2y$ into $y + y'$.

**6.34 Examples.**

1. Find the polar of the point $(5, 7)$ with regard to the ellipse
$$2x^2 + 3y^2 = 6.$$

2. Find the pole of the line $3x + 4y = 5$ with regard to the ellipse
$$3x^2 + 4y^2 = 5.$$

3. Find the poles of the lines $2x - y = 1$ and $x + 3y = 4$ with regard to the ellipse $x^2 + 4y^2 = 6$, and verify that the line joining the poles is the polar of the point of intersection of the lines.

4. Prove that the points $(4, 1)$, $(1, -1)$ are conjugate with regard to the ellipse $\qquad 2x^2 + 3y^2 = 5.$

5. Find the equation of a line through the point $(3, 2)$ and conjugate to the line $x + y = 1$ with regard to the ellipse
$$3x^2 + 4y^2 = 2.$$

6. Prove that, if the polar of a point is at right angles to the line joining the point to the centre of an ellipse, the point must lie on one of the axes.

**7.** Prove that, if $(x'', y'')$ is the pole of the normal at $(x', y')$, then

$$\frac{x'x''}{a^4} = -\frac{y'y''}{b^4} = \frac{1}{a^2-b^2}.$$

**8.** Prove that a directrix of an ellipse is the polar of the corresponding focus.

**6.4   Auxiliary circle and eccentric angle.** The circle described on the major axis $AA'$ of an ellipse as diameter is called the **auxiliary circle**.

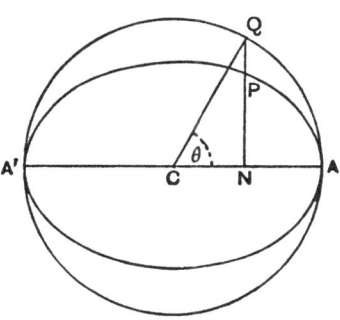

Let the ordinate $NP$ of any point $P$ on the ellipse be produced to meet the auxiliary circle in $Q$. Then the angle $ACQ$ is called the **eccentric angle** of the point $P$. As $P$ moves round the ellipse in the counter-clockwise sense in the figure starting from $A$ its eccentric angle increases from 0 to $2\pi$.

If $P$ is the point $(x, y)$ we have $x = CN = a \cos \theta$, then from the equation $\dfrac{x^2}{a^2}+\dfrac{y^2}{b^2}=1$, it follows that $y = \pm b \sin \theta$; but $y$ is positive or negative according as $\sin \theta$ is positive or negative, therefore the plus sign must be taken and the co-ordinates of any point on the ellipse are

$$x = a \cos \theta, \quad y = b \sin \theta.$$

This constitutes a parametric representation of the ellipse, the elimination of the parameter $\theta$ leading at once to the ordinary equation

$$\frac{x^2}{a^2}+\frac{y^2}{b^2}=1.$$

Since $NP = b \sin \theta$ and $NQ = a \sin \theta$, therefore $NP:NQ = b:a$; so that the ellipse may be derived from the auxiliary circle by reducing the ordinates in the circle in the same proportion.

**6.41 Chord joining the points whose eccentric angles
are $\theta$ and $\phi$.** The co-ordinates of the points being ($a \cos \theta$,
$b \sin \theta$) and ($a \cos \phi$, $b \sin \phi$), the equation of the line joining
them is
$$\frac{x - a \cos \theta}{a (\cos \theta - \cos \phi)} = \frac{y - b \sin \theta}{b (\sin \theta - \sin \phi)},$$

or $\dfrac{x - a \cos \theta}{2a \sin \frac{1}{2} (\theta + \phi) \sin \frac{1}{2} (\phi - \theta)} = \dfrac{y - b \sin \theta}{2b \sin \frac{1}{2} (\theta - \phi) \cos \frac{1}{2} (\theta + \phi)},$

which, on rejecting the common factor, reduces to

$$\frac{x}{a} \cos \tfrac{1}{2} (\theta + \phi) + \frac{y}{b} \sin \tfrac{1}{2} (\theta + \phi) = \cos \tfrac{1}{2} (\theta - \phi) \ \ ......(1).$$

The equation of the tangent at the point ($a \cos \theta$, $b \sin \theta$) is
found by making the second point ($a \cos \phi$, $b \sin \phi$) move up to
coincidence with the first. This gives

$$\frac{x}{a} \cos \theta + \frac{y}{b} \sin \theta = 1 \qquad ......(2).$$

This agrees with the form of equation of the tangent given
in 6.22 if we put $\quad x' = a \cos \theta, \quad y' = b \sin \theta.$

**6.42** Further, if $Q$, $Q'$ are the points whose eccentric angles are
$\alpha + \beta$ and $\alpha - \beta$, we find by writ-
ing $\theta = \alpha + \beta$ and $\phi = \alpha - \beta$ in
6.41 (1) that the equation of the
chord $QQ'$ is

$\dfrac{x}{a} \cos \alpha + \dfrac{y}{b} \sin \alpha = \cos \beta \quad ...(1).$

Now if we keep $\alpha$ fixed and
diminish $\beta$ the points $Q$, $Q'$ will
move up to coincidence at a

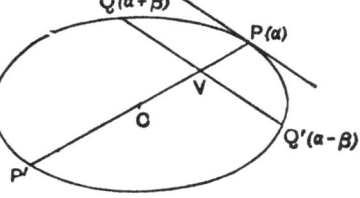

point $P$ whose eccentric angle is $\alpha$ and the chord will become the
tangent at $P$ and, by putting $\beta = 0$ in (1), the equation of this tangent
is
$$\frac{x}{a} \cos \alpha + \frac{y}{b} \sin \alpha = 1 \qquad ......(2).$$

By comparing (1) and (2) we see that for all values of $\beta$ the chord
joining the points whose eccentric angles are $\alpha + \beta$ and $\alpha - \beta$ is parallel
to the tangent at the point whose eccentric angle is $\alpha$.

**6.43    The normal at** $(a \cos \theta, \, b \sin \theta)$. The normal being at right angles to the tangent its equation is

$$\frac{a}{\cos \theta}\,(x - a \cos \theta) - \frac{b}{\sin \theta}\,(y - b \sin \theta) = 0,$$

or                    $ax \sec \theta - by \, \text{cosec}\, \theta = a^2 - b^2.$

**6.5    Locus of middle points of parallel chords. Conjugate diameters.** In the notation of 6.42, let $Q$, $Q'$ be the points whose eccentric angles are $\alpha + \beta$, $\alpha - \beta$, then the equation of $QQ'$ is

$$\frac{x}{a}\cos \alpha + \frac{y}{b}\sin \alpha = \cos \beta \dots(1),$$

and, if $P$ is the point whose eccentric angle is $\alpha$, $QQ'$ is parallel to the tangent at $P$ for all values of $\beta$.

Also the co-ordinates of the middle point $V$ of $QQ'$ are

$$x = \tfrac{1}{2}a \{\cos (\alpha + \beta) + \cos (\alpha - \beta)\},$$
$$y = \tfrac{1}{2}b \{\sin (\alpha + \beta) + \sin (\alpha - \beta)\},$$
or                    $x = a \cos \alpha \cos \beta,$
$$y = b \sin \alpha \cos \beta \qquad \qquad \dots\dots(2).$$

But the co-ordinates of $P$ are $a \cos \alpha$, $b \sin \alpha$, so that the co-ordinates of $V$ being proportional to those of $P$ the point $V$ lies on the radius $CP$.

Hence any diameter $PCP'$ (line through the centre) of an ellipse bisects all chords parallel to the tangents at its extremities.

Now let $PCP'$ be the line $y = mx$, and let the diameter $DCD'$ parallel to the tangent at $P$ be $y = m'x$.

Then $y = mx$ passes through $P$, whose co-ordinates are $a \cos \alpha$, $b \sin \alpha$, so that $m = \dfrac{b}{a}\tan \alpha$; and $y = m'x$ is parallel to

$$\frac{x}{a}\cos \alpha + \frac{y}{b}\sin \alpha = 1,$$

so that                    $m' = -\dfrac{b}{a}\cot \alpha.$

Therefore $$mm' = -\frac{b^2}{a^2} \qquad \ldots\ldots(3)$$

is the condition that the line $y = mx$ may bisect all chords parallel to $y = m'x$.

The symmetry of this relation between $m$ and $m'$ shows that it is also the condition that the diameter $DCD'$ may bisect all chords parallel to $PCP'$.

Such a pair of diameters are said to be **conjugate**.

**6.51 Another form of the condition for conjugate diameters.**
Let $x_1$, $y_1$ and $x_2$, $y_2$ be the co-ordinates of the points $P$ and $D$.

The equations of $CP$ and $CD$ may then be written

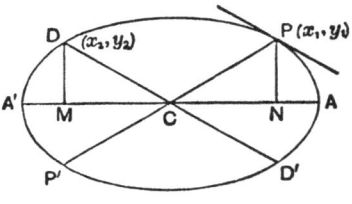

$$y = \frac{y_1}{x_1}\,x \text{ and } y = \frac{y_2}{x_2}\,x,$$

and the tangent at $P$ is

$$\frac{xx_1}{a^2} + \frac{yy_1}{b^2} = 1,$$

so the condition that $CD$ may be parallel to the tangent at $P$ is

$$\frac{y_2}{x_2} = -\frac{b^2}{a^2}\frac{x_1}{y_1},$$

which may be written $$\frac{x_1x_2}{a^2} + \frac{y_1y_2}{b^2} = 0 \qquad \ldots\ldots(1),$$

and the symmetry of this relation shows that it is also the condition that $CP$ may be parallel to the tangent at $D$.

By writing $m = y_1/x_1$ and $m' = y_2/x_2$ we see that this condition is identical with 6.5 (3), viz. $mm' = -b^2/a^2$.

**6.52 The eccentric angles of the ends of conjugate radii differ by a right angle.** If $\theta$, $\phi$ are the eccentric angles of the points $P$, $D$ of 6.51, we have

$$x_1 = a\cos\theta, \quad y_1 = b\sin\theta, \quad x_2 = a\cos\phi, \quad y_2 = b\sin\phi,$$

so that (1) gives $\cos\theta\cos\phi + \sin\theta\sin\phi = 0,$

or $$\cos(\theta - \phi) = 0.$$

Hence $$\theta \sim \phi = \tfrac{1}{2}\pi.$$

**6.53** **The sum of the squares of two conjugate radii is constant.** In the figure of **6.51** let $\theta$ be the eccentric angle of $P$ then $\theta + \frac{1}{2}\pi$ is that of $D$ and $\theta - \frac{1}{2}\pi$ that of $D'$.

Hence the co-ordinates of $D$ are

$$a \cos (\theta + \tfrac{1}{2}\pi), \quad b \sin (\theta + \tfrac{1}{2}\pi) \quad \text{or} \quad -a \sin \theta, \; b \cos \theta,$$

and those of $D'$ are $a \sin \theta, \; -b \cos \theta$.

Therefore we have

$$CP^2 = a^2 \cos^2 \theta + b^2 \sin^2 \theta$$

and

$$CD^2 = a^2 \sin^2 \theta + b^2 \cos^2 \theta,$$

whence

$$CP^2 + CD^2 = a^2 + b^2.$$

It is further easily seen from the figure that if $PN$, $DM$ are ordinates

$$NP : CM = b : a$$

and

$$CN : MD = a : b.$$

**6.54** **The area of the parallelogram formed by tangents at the ends of conjugate diameters is constant.** Let $PCP'$, $DCD'$ be the conjugate diameters, and let $F$ be the point where the normal at $P$ cuts $DCD'$. The area of the parallelogram formed by the four tangents is clearly $4PF.CD$.

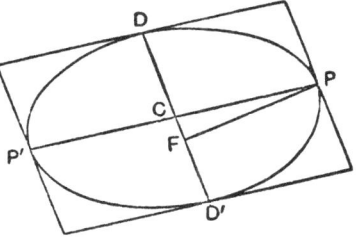

Now $PF$ is equal to the perpendicular from the centre to the tangent at $P$, whose equation is

$$\frac{x}{a} \cos \theta + \frac{y}{b} \sin \theta = 1,$$

so that $PF = \dfrac{1}{\sqrt{\left(\dfrac{\cos^2 \theta}{a^2} + \dfrac{\sin^2 \theta}{b^2}\right)}} = \dfrac{ab}{\sqrt{(a^2 \sin^2 \theta + b^2 \cos^2 \theta)}}$,

and from 6.53 $\quad CD = \sqrt{(a^2 \sin^2 \theta + b^2 \cos^2 \theta)}$.

Therefore $\qquad\qquad PF.CD = ab$

and the required area $= 4ab$.

**6.55** **Equiconjugate diameters.** From 6.53 it follows that $CP = CD$ if

$$a^2 \cos^2 \theta + b^2 \sin^2 \theta = a^2 \sin^2 \theta + b^2 \cos^2 \theta,$$

and this requires that $\theta =$ an odd multiple of $\frac{1}{4}\pi$.

Hence the diameters whose equations are $\dfrac{x}{a} \pm \dfrac{y}{b} = 0$ are conjugate diameters of equal length.

**6.56　Examples.** The following examples refer to the ellipse whose equation is
$$x^2/a^2 + y^2/b^2 = 1.$$

**1.** Show that the eccentric angle of one end of a latus rectum is $\cos^{-1} e$.

**2.** If $\alpha$ is the eccentric angle of a point $P$ on an ellipse, what are the co-ordinates of the corresponding point $Q$ on the auxiliary circle? Write down the equations of the tangents at $P$ and $Q$ and show that they intersect on the major axis.

**3.** Prove that, if $CP$, $CD$ are conjugate radii of an ellipse and $\omega$ the angle between them, then $\sin^2 \omega$ varies as $CP^{-2} + CD^{-2}$.

**4.** Prove that if a parallelogram is inscribed in an ellipse its sides are parallel to conjugate diameters.

**5.** Prove that if $QQ'$ is a chord of an ellipse parallel to the tangent at $P$ the eccentric angles of $Q$ and $Q'$ differ from the eccentric angle at $P$ by equal amounts.

**6.** Prove that the equation of the perpendicular bisector of the chord joining the points $Q$, $Q'$ whose eccentric angles are $\alpha + \beta$, $\alpha - \beta$ is
$$ax \sec \alpha - by \operatorname{cosec} \alpha = (a^2 - b^2) \cos \beta,$$
and deduce the equation of the normal at the point whose eccentric angle is $\alpha$.

**7.** Prove that, if the same chord passes through a focus, then
$$\cos \beta = \pm e \cos \alpha.$$

**8.** Prove that the equation of a focal chord parallel to the tangent at the point whose eccentric angle is $\alpha$ is
$$\frac{x}{a} \cos \alpha + \frac{y}{b} \sin \alpha = \pm e \cos \alpha.$$

**9.** Prove that, if $\alpha$ is variable and $\beta$ constant, the chord joining the points whose eccentric angles are $\alpha + \beta$ and $\alpha - \beta$ touches the ellipse
$$x^2/a^2 + y^2/b^2 = \cos^2 \beta,$$
and that the locus of the poles of the chord is
$$x^2/a^2 + y^2/b^2 = \sec^2 \beta.$$

**10.** Prove that the tangents at the points whose eccentric angles are $\alpha$, $\alpha + \frac{1}{2}\pi$ intersect on the ellipse
$$x^2/a^2 + y^2/b^2 = 2.$$

**11.** Prove that, if $P$, $Q$ are corresponding points on an ellipse and its auxiliary circle and the normals at $P$, $Q$ intersect in $R$, then

$$CR = a + b.$$

**12.** Prove that, if the line joining the ends of two equal conjugate radii of an ellipse passes through a focus the eccentricity is $1/\sqrt{2}$.

**18.** Prove that the chords that join the ends of conjugate radii all touch the ellipse $x^2/a^2 + y^2/b^2 = \frac{1}{2}$.

**6.6 Further geometrical properties of the ellipse.** Let the tangent at $P$ meet the major and minor axes in $T$, $T'$ and let perpendiculars from the foci $S$, $S'$ meet the tangent at $P$ in $Y$, $Y'$.

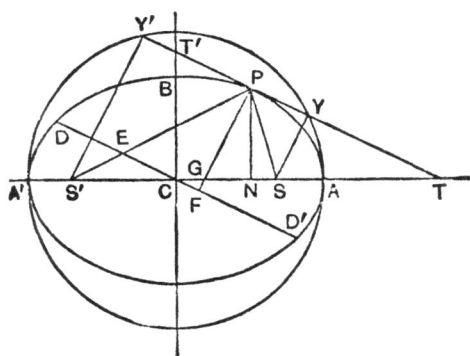

Let the diameter conjugate to $CP$ meet the normal $PG$ in $F$ and the focal distance $S'P$ in $E$.

Let $(x', y')$ be the co-ordinates of $P$, then the tangent is the line

$$\frac{xx'}{a^2} + \frac{yy'}{b^2} = 1,$$

and it meets the major axis $y = 0$ where $xx' = a^2$, i.e. in the point $T$ such that
$$CN . CT = CA^2 \qquad \qquad \ldots\ldots(1);$$
similarly it meets the minor axis $x = 0$ where $yy' = b^2$, i.e. in the point $T'$ such that
$$CT' . NP = CB^2 \qquad \qquad \ldots\ldots(2).$$

Again, if $\theta$ is the eccentric angle of $P$, the equation of the tangent may be written
$$bx \cos \theta + ay \sin \theta = ab \qquad \qquad \ldots\ldots(3)$$
and the equation of the perpendicular line through the focus $S$ $(ae, 0)$ is
$$ax \sin \theta - by \cos \theta = a^2 e \sin \theta;$$

whence by squaring and adding the last two equations it follows that the intersection $Y$ of the two lines lies on the locus

$$(x^2 + y^2)(b^2 \cos^2 \theta + a^2 \sin^2 \theta) = a^2 (b^2 + a^2 e^2 \sin^2 \theta).$$

But $b^2 + a^2 e^2 \sin^2 \theta = b^2 + (a^2 - b^2) \sin^2 \theta = b^2 \cos^2 \theta + a^2 \sin^2 \theta$;

therefore the point $Y$ lies on the circle

$$x^2 + y^2 = a^2 \qquad \qquad \ldots \ldots (4).$$

Similarly for the point $Y'$, so that the projections of the foci on any tangent lie on the auxiliary circle.

Again from 6.11    $SP = a - ex' = a(1 - e \cos \theta)$

and             $S'P = a + ex' = a(1 + e \cos \theta),$

so that           $SP . S'P = a^2 (1 - e^2 \cos^2 \theta)$

$$= a^2 - (a^2 - b^2) \cos^2 \theta$$

$$= a^2 \sin^2 \theta + b^2 \cos^2 \theta$$

$$= CD^2 \quad (6.53) \qquad \qquad \ldots \ldots (5).$$

Also, $SY$ being the perpendicular from $(ae, 0)$ to the line (3), we have

$$SY = \frac{ab(1 - e \cos \theta)}{\sqrt{(b^2 \cos^2 \theta + a^2 \sin^2 \theta)}} = \frac{b . SP}{CD} \cdot \left.\begin{array}{c} \\ \\ \\ \end{array}\right\}$$

Similarly      $S'Y' = \dfrac{b . S'P}{CD},$             $\ldots \ldots (6),$

and             $SY . S'Y' = b^2 \dfrac{SP . S'P}{CD^2} = b^2 \qquad \ldots \ldots (7).$

From (6) we see that $SY : SP = S'Y' : S'P \qquad \qquad \ldots \ldots (8),$

so that the right-angled triangles $SYP$, $S'Y'P$ are similar and the angles $SPY$, $S'PY'$ are equal, i.e. *the tangent to an ellipse is equally inclined to the focal distances.*

Since the triangles $PFE$, $S'Y'P$ are similar, we have

$$\frac{PF}{PE} = \frac{S'Y'}{S'P} = \frac{b}{CD},$$

from (6). But          $PF . CD = ab \quad (6.54),$

therefore               $PE = a \qquad \qquad \ldots \ldots (9).$

Again, the equation of the normal at $P$ is

$$\frac{x - x'}{\dfrac{x'}{a^2}} = \frac{y - y'}{\dfrac{y'}{b^2}} = -rp \quad (6.24) \qquad \qquad \ldots \ldots (10),$$

where $p$ is the central perpendicular on the tangent, and $r$ is the distance of the point $(x, y)$ from $P$ along the normal drawn inwards.

In the figure $p = PF$, and if we take $(x, y)$ to be $G$ we put $y = 0$ in the equation and find

$$rp = b^2 \quad \text{or} \quad PG.PF = b^2 \qquad \ldots\ldots(11).$$

By the same substitution $y = 0$, we also find that

$$x = x'\left(1 - \frac{b^2}{a^2}\right) = e^2x', \quad \text{or} \quad CG = e^2x';$$

whence it follows that

$$SG = CS - CG = ae - e^2x'$$
$$= e(a - ex') = eSP \quad (6.11) \quad \ldots\ldots(12),$$

and similarly $S'G = eS'P$; then from the fact that

$$SG : S'G = SP : S'P$$

we have a verification that $PG$ bisects the angle $S'PS$ which also follows from what precedes.

**6.7   The tangents at the ends of a chord of an ellipse intersect on the diameter that bisects the chord.** With the notation of 6.42, $Q$, $Q'$ and $P$ are the points whose eccentric angles are $\alpha + \beta$, $\alpha - \beta$ and $\alpha$. The equation of the chord $QQ'$ is, by 6.42,

$$\frac{x}{a}\cos\alpha + \frac{y}{b}\sin\alpha = \cos\beta \qquad \ldots\ldots(1),$$

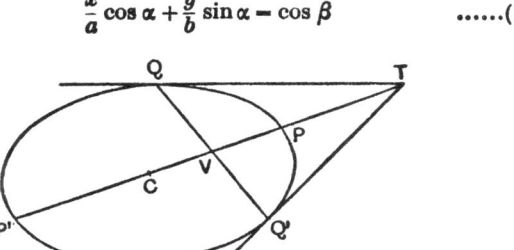

and by comparing this equation with that of the polar of a point $(x', y')$ we see that the pole $T$ of $QQ'$, or the intersection of the tangents at $Q$ and $Q'$, must be the point $(a\cos\alpha\sec\beta, b\sin\alpha\sec\beta)$ and this point clearly lies on the diameter through $P$ $(a\cos\alpha, b\sin\alpha)$ which bisects $QQ'$.

**6.71   Further geometrical properties of the ellipse.** In the figure of 6.7 we have seen that the co-ordinates of $T$ are $a\cos\alpha\sec\beta$, $b\sin\alpha\sec\beta$, so that

$$CT^2 = (a^2\cos^2\alpha + b^2\sin^2\alpha)\sec^2\beta,$$
or $$CT = CP\sec\beta \qquad \ldots\ldots(1).$$

Again from **6.5** (2) the co-ordinates of $V$ are $a \cos \alpha \cos \beta,\, b \sin \alpha \cos \beta$, so that
$$CV^2 = (a^2 \cos^2 \alpha + b^2 \sin^2 \alpha) \cos^2 \beta,$$
or
$$CV = CP \cos \beta \qquad\qquad ......(2);$$
whence we get
$$CV \cdot CT = CP^2 \qquad\qquad ......(3).$$
Also $Q$ is the point
$$\{a \cos (\alpha + \beta),\ b \sin (\alpha + \beta)\},$$
therefore
$$QV^2 = a^2 \{\cos (\alpha + \beta) - \cos \alpha \cos \beta\}^2 + b^2 \{\sin (\alpha + \beta) - \sin \alpha \cos \beta\}^2$$
$$= (a^2 \sin^2 \alpha + b^2 \cos^2 \alpha) \sin^2 \beta$$
$$= CD^2 \sin^2 \beta,$$
but
$$PV \cdot VP' = CP^2 - CV^2 = CP^2 \sin^2 \beta;$$
and from these results we deduce that
$$QV^2 : PV \cdot VP' = CD^2 : CP^2,$$
which is a theorem well known to students of geometrical conics.

**6.8  Segments of chords.** We shall now apply to the ellipse the method of **4.3**.

Let $O$ be a point $(x', y')$ then the equation of a line through $O$ making an angle $\theta$ with the major axis is, by **2.43**,
$$\frac{x - x'}{\cos \theta} = \frac{y - y'}{\sin \theta} = r \qquad\qquad ......(1).$$

By writing $x = x' + r \cos \theta$ and $y = y' + r \sin \theta$, we see that the line meets the ellipse when
$$\frac{(x' + r \cos \theta)^2}{a^2} + \frac{(y' + r \sin \theta)^2}{b^2} = 1,$$
or $r^2 \left( \dfrac{\cos^2 \theta}{a^2} + \dfrac{\sin^2 \theta}{b^2} \right) + 2r \left( \dfrac{x' \cos \theta}{a^2} + \dfrac{y' \sin \theta}{b^2} \right) + \dfrac{x'^2}{a^2} + \dfrac{y'^2}{b^2} - 1 = 0 \ ...(2).$

If $Q$ and $Q'$ are the points in which the line meets the ellipse then $OQ$, $OQ'$ are the roots of this quadratic in $r$.

The first deduction that we make from (2) is that if $O$ is the middle point of the chord $QQ'$, then the roots of (2) are equal and opposite in sign. This requires that
$$\frac{x'}{a^2} \cos \theta + \frac{y'}{b^2} \sin \theta = 0,$$
or that $O$ lies on the line
$$y = -\frac{b^2 x}{a^2} \cot \theta,$$
which is therefore the locus of the middle points of chords of the ellipse parallel to the line $y = x \tan \theta$.

This result accords with **6.5** (3).

**6.81** The second deduction made from the quadratic (2) is that the product of the roots

$$OQ.OQ' = \left(\frac{x'^2}{a^2} + \frac{y'^2}{b^2} - 1\right) \Big/ \left(\frac{\cos^2 \theta}{a^2} + \frac{\sin^2 \theta}{b^2}\right) \quad \ldots\ldots(3).$$

Similarly for a chord $ORR'$ making an angle $\theta'$ with the major axis

$$OR.OR' = \left(\frac{x'^2}{a^2} + \frac{y'^2}{b^2} - 1\right) \Big/ \left(\frac{\cos^2 \theta'}{a^2} + \frac{\sin^2 \theta'}{b^2}\right) \quad \ldots\ldots(4),$$

whence we get

$$\frac{OQ.OQ'}{OR.OR'} = \left(\frac{\cos^2 \theta'}{a^2} + \frac{\sin^2 \theta'}{b^2}\right) \Big/ \left(\frac{\cos^2 \theta}{a^2} + \frac{\sin^2 \theta}{b^2}\right) \quad \ldots\ldots(5).$$

We now observe that this ratio depends only on the directions of the chords and not on the position of the point $O$. Hence, if through any other point $o$ chords $oqq'$, $orr'$ are drawn parallel respectively to $OQQ'$ and $ORR'$ we have

$$\frac{OQ.OQ'}{OR.OR'} = \frac{oq.oq'}{or.or'}.$$

In particular, if we take $o$ at the centre and $PCP'$, $DCD'$ are diameters parallel to $OQQ'$, $ORR'$ we have

$$\frac{OQ.OQ'}{OR.OR'} = \frac{CP^2}{CD^2}.$$

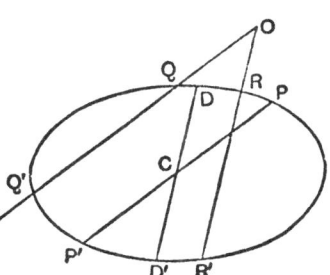

**6.82** Intersections of a circle and an ellipse. If a circle cuts an ellipse in the points $Q$, $Q'$, $R$, $R'$, and the chords $QQ'$, $RR'$ meet in $O$, then from the property of the circle $OQ.OQ' = OR.OR'$; and from 6.81 it follows that the radii of the ellipse parallel to $QQ'$ and $RR'$ must be equal. But equal radii of an ellipse are equally inclined to the axes. Hence the chords of intersection of a circle and an ellipse are equally inclined to the axes.

**6.83** The student may apply the method of 6.8 to the parabola and make similar deductions, including the theorem that the ratio of the rectangles under the segments of two intersecting chords of a parabola is the same as the ratio of the lengths of parallel focal chords.

**6.9 Examples.**

(i) *Prove that the distance from the centre of a normal to the ellipse* $x^2/a^2 + y^2/b^2 = 1$, *is never greater than* $a - b$.

The equation of the normal at $(a \cos \alpha, b \sin \alpha)$ is

$$ax \sec \alpha - by \csc \alpha = a^2 - b^2 \quad (6.43).$$

The length of the perpendicular from the centre is therefore

$$\frac{a^2 - b^2}{\sqrt{(a^2 \sec^2 \alpha + b^2 \operatorname{cosec}^2 \alpha)}}.$$

This is greatest when the denominator is least. But

$$a^2 \sec^2 \alpha + b^2 \operatorname{cosec}^2 \alpha = a^2 + b^2 + a^2 \tan^2 \alpha + b^2 \cot^2 \alpha$$
$$= (a + b)^2 + (a \tan \alpha - b \cot \alpha)^2.$$

This has its least value $(a + b)^2$ when $a \tan \alpha = b \cot \alpha$, and the perpendicular has then its greatest value $a - b$.

(ii) *A normal chord of the ellipse $x^2/a^2 + y^2/b^2 = 1$ subtends a right angle at the centre. Prove that the pole of the chord lies on the ellipse*

$$x^2/a^4 + y^2/b^4 = 1/a^2 + 1/b^2.$$

The equation of the normal at $(a \cos \alpha, b \sin \alpha)$ is

$$ax \sec \alpha - by \operatorname{cosec} \alpha = a^2 - b^2 \qquad \ldots\ldots(1);$$

comparing this with the equation of the polar of $(x', y')$, viz.

$$\frac{xx'}{a^2} + \frac{yy'}{b^2} = 1,$$

we see that $\qquad x' = \dfrac{a^3 \sec \alpha}{a^2 - b^2}$ and $y' = -\dfrac{b^3 \operatorname{cosec} \alpha}{a^2 - b^2} \qquad \ldots\ldots(2).$

To express the condition that the intersections of the line (1) and the ellipse subtend a right angle at the centre, we form an equation to represent the lines joining the centre to the intersections. This is achieved by making a homogeneous equation from (1) and the equation of the ellipse. Such an equation is

$$\frac{x^2}{a^2} + \frac{y^2}{b^2} = \frac{(ax \sec \alpha - by \operatorname{cosec} \alpha)^2}{(a^2 - b^2)^2}.$$

This equation is clearly satisfied by the points common to the line (1) and the ellipse and being homogeneous it represents straight lines through the origin (2.7).

Further, it is an equation of the second degree and by 2.71 the condition that the lines it represents may be at right angles is that the sum of the coefficients of $x^2$ and $y^2$ should be zero; i.e. we must have

$$\frac{1}{a^2} + \frac{1}{b^2} = \frac{a^2 \sec^2 \alpha}{(a^2 - b^2)^2} + \frac{b^2 \operatorname{cosec}^2 \alpha}{(a^2 - b^2)^2},$$

so that, making use of (2), we have

$$\frac{1}{a^2} + \frac{1}{b^2} = \frac{x'^2}{a^4} + \frac{y'^2}{b^4}.$$

Therefore the pole $(x', y')$ of the normal chord lies on the ellipse

$$\frac{x^2}{a^4} + \frac{y^2}{b^4} = \frac{1}{a^2} + \frac{1}{b^2}.$$

**6.91 Harder examples.** In the following examples on the ellipse it may be assumed that its equation is $x^2/a^2 + y^2/b^2 = 1$, and that $C$ is its centre and $S$, $S'$ its foci.

1. Prove that, if the tangent and normal at a point $P$ on an ellipse meet the major axis in $T$, $G$, then the tangent from either end of the minor axis to the circle $TPG$ is equal in length to half the major axis.

2. Show that if $x'$, $y'$ are the co-ordinates of a point of intersection of the ellipses $x^2/a^2 + y^2/b^2 = 1$ and $x^2/a'^2 + y^2/b'^2 = 1$, the equations of their common tangents are $\pm\dfrac{xx'}{aa'} \pm \dfrac{yy'}{bb'} = 1$.

3. The normal at any point $P$ of the ellipse meets the axis in $G$; a point $Q$ is taken in the tangent so that $PQ = \lambda.PG$, where $\lambda$ is constant; prove that the locus of $Q$ is the ellipse

$$\frac{x^2}{a^2} + \frac{y^2}{b^2} = \frac{a^2 + \lambda^2 b^2}{a^2}.$$

4. Prove that the line $\dfrac{ax}{k^2-1} + \dfrac{by}{2k} + \dfrac{a^2-b^2}{k^2+1} = 0$ is a normal to the ellipse $x^2/a^2 + y^2/b^2 = 1$ for all values of $k$.

5. Prove that the foot of the focal perpendicular on the normal at any point of an ellipse is at a distance from the centre equal to the difference between the semi-major axis and the focal radius vector to the point at which the normal is drawn.

6. Prove that the locus of the poles of normal chords of the ellipse is the curve

$$\frac{a^6}{x^2} + \frac{b^6}{y^2} = (a^2 - b^2)^2.$$

7. $P$ is a point $(x', y')$ on the ellipse and $PS$, $PS'$ meet the curve again in $Q$, $R$. Prove that the equation of $QR$ is

$$\frac{xx'}{a^2} + \frac{yy'}{b^2}\frac{1+e^2}{1-e^2} + 1 = 0,$$

where $e$ is the eccentricity.

8. Show that two parallel tangents to an ellipse are met by any other tangent in points which lie on conjugate diameters.

9. Prove that if $CP$, $CD$ be any two conjugate semi-diameters of an ellipse and $PF$ be drawn perpendicular to $CD$ and produced both ways to $E$, $E'$ so that $PE = PE' = CD$, then $CE \cdot CE' = CS^2$ where $S$ is a focus.

10. Tangents are drawn from points on the ellipse to the circle $x^2 + y^2 = r^2$, show that the chords of contact touch the ellipse

$$a^2x^2 + b^2y^2 = r^4.$$

11. Tangents $TP$, $TQ'$ are drawn to an ellipse so that $SP$, $S'Q$ are parallel. Prove that $CT$ is parallel to $SP$ or $S'Q$.

12. The normal to an ellipse at a point $P$ passes through one end of the minor axis, and $CD$ is the semidiameter conjugate to $CP$. The perpendicular from $C$ to $CD$ meets the auxiliary circle in $E$. Prove that $DE$ is equal to half the distance between the directrices.

13. Prove that, if $(x', y')$ be the middle point of a chord of the ellipse, the equation of the chord is

$$\frac{xx'}{a^2} + \frac{yy'}{b^2} = \frac{x'^2}{a^2} + \frac{y'^2}{b^2}.$$

14. If $Q$ be the pole of the chord of the ellipse which is normal at a point $P$, and if $CR$ drawn through the centre $C$ perpendicular to $CQ$ meet the normal at $P$ in $R$, prove that the locus of $R$ is

$$x^2/b^6 + y^2/a^6 = 1/a^2b^2.$$

15. Prove that, if the point $P$ lies on the ellipse $x^2/a'^2 + y^2/b'^2 = 1$, its polar with regard to the ellipse $x^2/a^2 + y^2/b^2 = 1$ touches the ellipse

$$a'^2x^2/a^4 + b'^2y^2/b^4 = 1.$$

16. Show that the polar with regard to the ellipse $x^2/a^2 + y^2/b^2 = 1$ of a point on the circle $x^2 + y^2 = c^2$ touches the ellipse $x^2/a^4 + y^2/b^4 = 1/c^2$.

17. Prove that, if the tangents to an ellipse at $(x_1, y_1)$ and $(x_2, y_2)$ meet at $(x, y)$ and the normals at $(\xi, \eta)$, then $a^2\xi = e^2xx_1x_2$ and $b^2\eta = -e^2yy_1y_2$, where $e$ is the eccentricity.

18. Show that, if $(x, y)$ is the middle point of a chord of the ellipse, and the tangents at the ends of the chord intersect in $(x', y')$ and the normals in $(x'', y'')$, then

$$\frac{a^2x''}{x'} + \frac{b^2y''}{y'} = (a^2 - b^2)\left(\frac{xx'}{a^2} - \frac{yy'}{b^2}\right).$$

**19.** Prove that, if the normals at two points $P$, $Q$ on an ellipse intersect on the diameter that bisects $PQ$, then the two normals are at right angles.

**20.** Prove that if a chord of the ellipse subtends a right angle at the centre then it touches the circle

$$(x^2 + y^2)(a^2 + b^2) = a^2b^2.$$

**21.** The locus of middle points of chords of the ellipse which subtend a right angle at its centre is

$$\frac{x^2}{a^4} + \frac{y^2}{b^4} = \frac{a^2 + b^2}{a^2b^2} \cdot \left(\frac{x^2}{a^2} + \frac{y^2}{b^2}\right)^2.$$

**22.** Show that the tangents at the extremities of all chords of the ellipse which subtend a right angle at the centre, intersect on the ellipse

$$\frac{x^2}{a^4} + \frac{y^2}{b^4} = \frac{1}{a^2} + \frac{1}{b^2}.$$

**23.** If $P$ be any point on the ellipse whose axes are $AA'$, $BB'$; and if the parallel lines $AP$, $BQ$ be drawn, $Q$ being on the ellipse; $Q$ will be one extremity of the diameter conjugate to that drawn from $P$.

**24.** From any point $T$ on one of the equiconjugate diameters of a conic whose centre is $O$ tangents $TP$, $TQ$ are drawn to the conic. Show that $O$, $P$, $Q$, $T$ are concyclic.

**25.** Prove that, if two conjugate radii of an ellipse cut the director circle in $T$, $T'$, then $TT'$ touches the ellipse.

**26.** If $PSP'$, $QSQ'$ be two focal chords and if $PQ$ be parallel to the major axis, show that $P'Q'$ bisects the distance between $S$ and the nearer directrix.

**27.** Tangents are drawn at the extremities of conjugate diameters of an ellipse, and meet in $O$. Prove that the perpendicular from $O$ on the focal radius to a point of contact is half the minor axis.

**28.** Show that the area of the rectangle formed by two parallel tangents and the corresponding normals to an ellipse is never greater than half the square on the line joining the foci.

**29.** Prove that the angle between the normal and the central radius at a point on an ellipse is greatest when the point is the end of one of the equiconjugate diameters.

**80.** $P$, $Q$ are two points on the ellipse, and $PS$, $QS'$ intersect on the curve, prove that the locus of the pole of $PQ$ is

$$\frac{x^2}{a^2} + \frac{b^2 y^2}{(2a^2 - b^2)^2} = 1.$$

**81.** Two conjugate diameters of an ellipse meet a fixed straight line $lx + my = 1$ in $P$, $Q$, and the straight lines through $P$, $Q$ perpendicular to these diameters intersect in $R$; prove that the locus of $R$ is the straight line $\quad a^2 lx + b^2 my = a^2 + b^2$.

**82.** Prove that if $\alpha$, $\beta$ are the eccentric angles of two points $P$, $P'$ on an ellipse such that the focal distances $SP$, $S'P'$ are parallel, then

$$\tan \tfrac{1}{2}\alpha : \tan \tfrac{1}{2}\beta = 1 \pm e : 1 \mp e,$$

and $PP'$ touches the ellipse

$$x^2/a^4 + y^2/b^4 = 1/a^2.$$

**33.** Prove that the square of the sum of two conjugate radii is greatest when the radii are equal.

**34.** Prove that, if on the inward normal to an ellipse at $P$ a length $PQ$ be taken equal to the conjugate radius $CD$, the locus of $Q$ is a circle of radius $a - b$.

**35.** Prove that, if $P$, $Q$ are corresponding points on an ellipse and its auxiliary circle, and the normal at $P$ to the ellipse meets the normal at $Q$ to the circle in $R$, then the locus of $R$ is a circle of radius $a + b$.

**36.** Prove that, if lines drawn from any point on an ellipse to the ends of a diameter $PCP'$ meet the conjugate diameter $DCD'$ in $M$, $M'$, then $CM \cdot CM' = CD^2$.

**37.** Prove that, if an ellipse slides between two straight lines at right angles to one another, the locus of its centre is a circle.

# CHAPTER VII

# THE HYPERBOLA

**7.1 DEF.** A **hyperbola** is the locus of a point which moves so that the ratio of its distance from a fixed point to its distance from a fixed straight line is a constant, greater than unity.

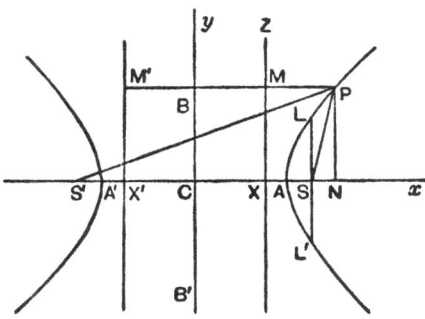

The fixed point is called the *focus*, the fixed line the *directrix* and the constant ratio the *eccentricity*. The latter is denoted by *e*.

*To find the equation of a hyperbola.*

Let $S$ be the focus and $ZX$ the directrix. Draw $SX$ perpendicular to the directrix and divide it internally and externally at $A$ and $A'$ in the ratio $e:1$, so that $A$ and $A'$ are points on the locus.

Let $C$ be the middle point of $AA'$ and let $A'A = 2a$.

Then we have

$$e = \frac{A'S}{A'X} = \frac{AS}{XA} = \frac{A'S + AS}{A'X + XA} = \frac{A'S - AS}{A'X - XA} = \frac{2CS}{2CA} = \frac{2CA}{2CX}.$$

Therefore $CS = eCA = ea$ and $CX = CA/e = a/e$ ......(1).

Take $C$ as origin and the $x$-axis along $CX$, and let $x$, $y$ be the co-ordinates of any point $P$ on the locus, $NP$ the ordinate of $P$ and $PM$ the perpendicular to the directrix.

Then $$SP^2 = e^2PM^2 = e^2NX^2,$$

therefore $$SN^2 + NP^2 = e^2(CN - CX)^2;$$

but $$SN = CN - CS = x - ae,$$

therefore $$(x - ae)^2 + y^2 = e^2\left(x - \frac{a}{e}\right)^2 = (ex - a)^2,$$

or $$x^2(e^2 - 1) - y^2 = a^2(e^2 - 1),$$

and by dividing by $a^2(e^2 - 1)$ this gives

$$\frac{x^2}{a^2} - \frac{y^2}{a^2(e^2 - 1)} = 1 \qquad \ldots\ldots(2).$$

Since $e$ is greater than unity $a^2(e^2 - 1)$ is a positive number and if we write $b^2$ for $a^2(e^2 - 1)$ the equation becomes

$$\frac{x^2}{a^2} - \frac{y^2}{b^2} = 1 \qquad \ldots\ldots(3),$$

and this is the standard form of equation of a hyperbola.

From (3) we get $y = \pm\dfrac{b}{a}\sqrt{(x^2 - a^2)}$, so that for real values of $y$ we must have $x$ numerically greater than $a$. Hence no part of the curve lies between $x = -a$ and $x = a$ and outside these limits there are two equal values of $y$ opposite in sign corresponding to every value of $x$, and as $x$ increases indefinitely so does $y$. In fact since the sign of $x$ or $y$ may be changed in (3) without altering the equation the curve is symmetrical about both axes.

$A'A$ is called the *transverse axis* and the double ordinate $LSL'$ through the focus is called the latus rectum.

Since $$SL = eSX = e(CS - CX) = a(e^2 - 1) = b^2/a \quad \ldots\ldots(4),$$

therefore $b^2/a$ or $a(e^2 - 1)$ is the length of the semi-latus rectum.

If on the $y$-axis we mark off lengths $B'C = CB = b$, then $B'CB$ may be called the *conjugate* axis, but this does not of course imply that $B$, $B'$ are points on the curve.

We note that since $e$ may have any value greater than unity and $b^2 = a^2(e^2 - 1)$, therefore $b$ may be longer than $a$, and for this reason the terms major and minor axes are not suitable though the former is sometimes used for $AA'$.

**7.11 The second focus and directrix.** It is evident from (8) that if $(x, y)$ lies on the curve so does $(-x, -y)$ so that every chord through the origin $C$ is bisected at $C$. $C$ is called the centre of the curve. It follows that if on $CA'$ we take points $S'$ and $X'$ such that $CS' = eCA'$ and $CX' = CA'/e$ and describe a hyperbola with $S'$ as focus and directrix $Z'X'$ parallel to $ZX$ and $S'P = ePM'$ it will be the same hyperbola as that obtained with $S$ as focus and $ZX$ as directrix. $S'$ and $Z'X'$ are therefore the second focus and directrix.

*SP and S'P are called the focal distances and their difference is equal to the transverse axis.*

For $\qquad SP = eMP = eXN = e\,(CN - CX) = ex - a$

and $\qquad S'P = eM'P = eX'N = e\,(CN + X'C) = ex + a,$

so that $\qquad\qquad\qquad S'P - SP = 2a.$

**7.12** In the special case in which the focus $S$ lies on the directrix $ZX$, say at $X$, it is clear by similar triangles that all points $P$, for which $\dfrac{SP}{PM} = $ a definite constant $e$ greater than unity, lie on one or other of two fixed straight lines $KSL$, $K'SL'$. Such a pair of lines is the degenerate form of hyperbola obtained when a plane passes through the vertex of a cone and cuts it in two straight lines (see 5.1).

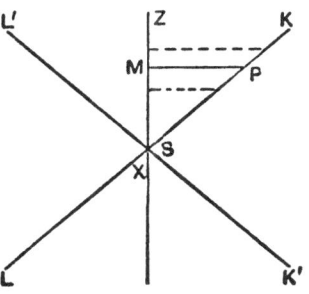

**7.2** Since the equation of a hyperbola only differs from that of an ellipse in the sign of $b^2$ most of the theorems proved for an ellipse have their counterparts for a hyperbola and in the proofs it is only necessary to change the sign of $b^2$. Thus taking the hyperbola to be $x^2/a^2 - y^2/b^2 = 1$ we have the following results.

(i) The lines $\qquad y = mx \pm \sqrt{(a^2m^2 - b^2)}$

are tangents to the curve (6.2). But these tangents are not real if $a^2m^2 - b^2$ is negative, i.e. if $|m| < b/a$, or if the lines make with the major axis an angle less than $\tan^{-1}(b/a)$.

(ii) The equation of the *director circle* is

$$x^2 + y^2 = a^2 - b^2 \quad (6.21)$$

and the circle only exists when $b \not> a$, and is a point at the centre when $b = a$.

(iii) The tangent at $(x', y')$ is

$$xx'/a^2 - yy'/b^2 = 1 \quad (6.22).$$

(iv) The line $lx + my + n = 0$ touches the hyperbola if

$$a^2l^2 - b^2m^2 = n^2,$$

and the line $x \cos \alpha + y \sin \alpha = p$ is a tangent if

$$a^2 \cos^2 \alpha - b^2 \sin^2 \alpha = p^2 \quad (6.23).$$

(v) The perpendicular distance $p$ of the tangent from the centre is given by

$$\frac{1}{p^2} = \frac{x'^2}{a^4} + \frac{y'^2}{b^4} \quad (6.231).$$

(vi) The normal at $(x', y')$ is

$$\frac{x - x'}{\dfrac{x'}{a^2}} = \frac{y - y'}{-\dfrac{y'}{b^2}} = \pm rp \quad (6.24),$$

where (by considering a point $(x', y')$ in the positive quadrant) we see that the negative sign of the $\pm$ must be taken if $(x, y)$ is on the normal drawn away from the transverse axis and the positive sign if on the normal drawn towards the axis.

(vii) The polar of $(x', y')$ is

$$xx'/a^2 - yy'/b^2 = 1 \quad (6.31).$$

(viii) The points $(x', y')$ and $(x'', y'')$ are conjugate if

$$x'x''/a^2 - y'y''/b^2 = 1,$$

and the lines $lx + my + n = 0$ and $l'x + m'y + n' = 0$ are conjugate if $\qquad a^2ll' - b^2mm' = nn' \quad (6.32).$

**7.3 Parametric representation of the hyperbola.** The student who is familiar with hyperbolic functions will realise at once that the equation of the hyperbola is satisfied by the substitutions $x = a \cosh \theta$, $y = b \sinh \theta$, and that there is then a close analogy between the use of this parameter and the eccentric angle in the ellipse.

Another substitution that may be used is

$$x = a \sec \theta, \quad y = b \tan \theta \qquad \ldots\ldots(1),$$

since substitution in the equation $x^2/a^2 - y^2/b^2 = 1$ leads to the identity
$$\sec^2 \theta - \tan^2 \theta = 1.$$

It then follows from **7.2** (iii) that the tangent at $(a \sec \theta, b \tan \theta)$ is

$$\frac{x}{a} \sec \theta - \frac{y}{b} \tan \theta = 1 \qquad \ldots\ldots(2).$$

The equation of the normal at the same point is therefore

$$\frac{a}{\sec \theta}(x - a \sec \theta) + \frac{b}{\tan \theta}(y - b \tan \theta) = 0$$

or
$$ax \cos \theta + by \cot \theta = a^2 + b^2 \qquad \ldots\ldots(3).$$

**7.31 Tracing the curve.** Taking the parametric values $x = a \sec \theta$, $y = b \tan \theta$, let $\theta$ increase continuously from $0$ to $2\pi$.

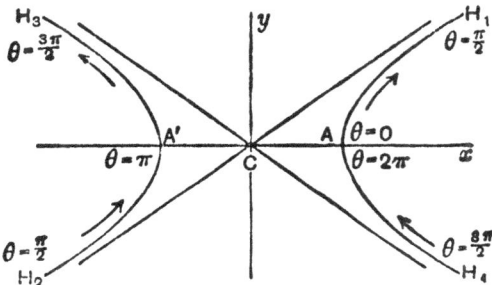

When $\theta = 0$ we have $x = a$, $y = 0$, i.e. the point $A$ on the curve; as $\theta$ increases from 0 to $\frac{1}{2}\pi$, $x$ and $y$ both increase to infinity positively and we have an infinite branch $AH_1$ in the first quadrant.

As $\theta$ increases past $\frac{1}{2}\pi$, $x$ and $y$ both change sign and decrease numerically so that we get a branch $H_2A'$ in the third quadrant arriving at $A'(-a, 0)$ when $\theta = \pi$. As $\theta$ increases past $\pi$, $y$ becomes positive and $x$ and $y$ increase numerically again to infinity as $\theta$ tends to $\frac{1}{2}\pi$ giving a branch $A'H_3$ in the second quadrant. Finally as $\theta$ increases past $\frac{3}{2}\pi$, $x$ becomes positive and $y$ negative and both decrease numerically, so that we get a branch $H_4A$, in the fourth quadrant, ending at $A(a, 0)$ when $\theta = 2\pi$.

**7.32   Asymptotes.** We have seen in **7.3** that the tangent at $(a \sec \theta, b \tan \theta)$ is

$$\frac{x}{a} \sec \theta - \frac{y}{b} \tan \theta = 1.$$

This can also be written

$$\frac{x}{a} - \frac{y}{b} \sin \theta = \cos \theta.$$

If we put $\theta = \frac{1}{2}\pi$, the equation of the tangent becomes

$$\frac{x}{a} - \frac{y}{b} = 0.$$

This is a line through the origin $C$, and from **7.31** it follows that it is the limiting position of the tangent to the curve as the point of contact moves to an infinite distance along either of the branches $AH_1$ or $A'H_2$.

Similarly, if we put $\theta = \frac{3}{2}\pi$, the equation of the tangent becomes

$$\frac{x}{a} + \frac{y}{b} = 0,$$

which is also a line through the origin and the limiting position of the tangent to the curve as the point of contact moves to an infinite distance along either of the branches $A'H_3$ or $AH_4$.

The lines

$$\frac{x}{a} - \frac{y}{b} = 0 \quad \text{and} \quad \frac{x}{a} + \frac{y}{b} = 0,$$

or, in one equation,

$$\frac{x^2}{a^2} - \frac{y^2}{b^2} = 0,$$

are called the **asymptotes** of the hyperbola. They are tangents that pass through the centre, their points of contact being at an infinite distance.

**7.33   The conjugate hyperbola.** Hyperbolas are said to be conjugate when the transverse and conjugate axes of one are the conjugate and transverse axes of the other. Thus the hyperbola

$$\frac{y^2}{b^2} - \frac{x^2}{a^2} = 1 \qquad \qquad \ldots\ldots(1)$$

is conjugate to the hyperbola

$$\frac{x^2}{a^2} - \frac{y^2}{b^2} = 1 \qquad \qquad \ldots\ldots(2);$$

for if $AA'$ and $BB'$ are the transverse and conjugate axes of the latter curve, the former meets the axis of $y$ in the real points $x = 0$, $y = \pm b$, i.e. in $B$ and $B'$ so that $BB'$ and $AA'$ are the transverse and conjugate axes of the former.

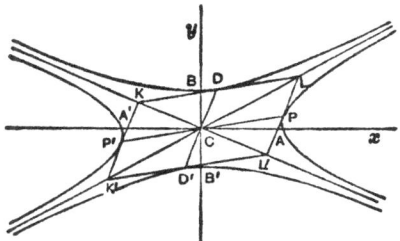

Both curves have the same asymptotes, namely the lines

$$\frac{x^2}{a^2} - \frac{y^2}{b^2} = 0.$$

The parametric equation of the conjugate hyperbola is

$$x = a \tan \theta, \quad y = b \sec \theta,$$

since these values of $x$ and $y$ satisfy (1).

**7.4 Segments of chords. Conjugate diameters.** The results of **6.8** may be adapted to the hyperbola by changing the sign of $b^2$, so that the geometrical theorems of **6.81** and **6.82** are true for a hyperbola as well as for an ellipse; and we also have the theorem that the locus of the middle points of chords of the hyperbola parallel to the line $y = x \tan \theta$ is the line $y = \frac{b^2}{a^2} \cot \theta$. So that we may say that the lines $y = mx$ and $y = m'x$ are conjugate diameters of the hyperbola, meaning that each bisects chords parallel to the other, if $mm' = b^2/a^2$.

**7.41 Conjugate diameters.** As in **6.5** diameters are conjugate when one is parallel to the tangents at the ends of the other.

Let a diameter $PCP'$ of the hyperbola

$$\frac{x^2}{a^2} - \frac{y^2}{b^2} = 1 \qquad \qquad \ldots\ldots(1)$$

meet the curve in points $P$, $P'$ and let $P$ be the point ($a \sec \theta$, $b \tan \theta$). The equation of the tangent at $P$ is

$$\frac{x}{a} \sec \theta - \frac{y}{b} \tan \theta = 1 \quad (7.3)$$

and the parallel line through the centre is therefore

$$\frac{x}{a} = \frac{y}{b} \sin \theta \qquad \qquad \ldots \ldots (2).$$

This line meets the curve (1) where

$$\frac{y^2}{b^2}(\sin^2 \theta - 1) = 1, \quad \text{or} \quad \frac{y^2}{b^2} = -\sec^2 \theta,$$

which does not give real values for $y$. So that the diameter conjugate to $PCP'$ does not meet the hyperbola (1) in real points.

But the line (2) meets the conjugate hyperbola

$$\frac{x^2}{a^2} - \frac{y^2}{b^2} = -1 \qquad \qquad \ldots \ldots (8)$$

where $\quad \frac{y^2}{b^2}(\sin^2 \theta - 1) = -1 \quad$ or $\quad y = \pm b \sec \theta$

and therefore $\qquad \qquad x = \pm a \tan \theta.$

We have therefore proved that if $PCP'$ is a diameter of (1) meeting it in real points ($\pm a \sec \theta$, $\pm b \tan \theta$), then the diameter parallel to the tangents at $P$ and $P'$ does not meet (1) in real points, but does meet the conjugate hyperbola (8) in real points $D$, $D'$ whose co-ordinates are

$$x = \pm a \tan \theta, \quad y = \pm b \sec \theta.$$

In order to avoid the consideration of imaginary points it is therefore convenient to speak of $PCP'$ and $DCD'$ as conjugate diameters, although $P$, $P'$ are points on the hyperbola (1) and $D$, $D'$ are points on the conjugate hyperbola (8).

**7.42**   $CP^2 - CD^2 = a^2 - b^2$. For, in the notation of **7.41**,

$$CP^2 = a^2 \sec^2 \theta + b^2 \tan^2 \theta$$

and $\qquad \qquad CD^2 = a^2 \tan^2 \theta + b^2 \sec^2 \theta,$

so that $\qquad \qquad CP^2 - CD^2 = a^2 - b^2.$

**7.43** *The vertices of the parallelogram formed by drawing tangents at the ends of two conjugate diameters lie on the asymptotes and its area is 4ab.*

Referring to the figure of **7.33**, let $PCP'$ and $DCD'$ be conjugate diameters and let the tangents at their extremities form the parallelogram $LL'K'K$.

Let $P$ be the point $(a \sec \theta, b \tan \theta)$, then by **7.41** the co-ordinates of $D$ are $a \tan \theta$, $b \sec \theta$. Therefore the middle point of $PD$ has co-ordinates $x = \frac{1}{2}a (\sec \theta + \tan \theta)$, $y = \frac{1}{2}b (\tan \theta + \sec \theta)$,

and lies on the line

$$\frac{x}{a} - \frac{y}{b} = 0,$$

i.e. on an asymptote.

But $PD$ is a diagonal of the parallelogram $CPLD$, and since its middle point lies on an asymptote, therefore the other diagonal $CL$ coincides in direction with the asymptote. In the same way it may be proved that the other vertices $L'$, $K'$, $K$ also lie on the asymptotes.

Again the area of the parallelogram $LL'K'K$

$$= 4CD \times \text{perpendicular from } C \text{ to } LL'.$$

But the equation of $LPL'$ is

$$\frac{x}{a} \sec \theta - \frac{y}{b} \tan \theta = 1,$$

so that the perpendicular from $C$

$$= \frac{1}{\sqrt{\left(\dfrac{\sec^2 \theta}{a^2} + \dfrac{\tan^2 \theta}{b^2}\right)}},$$

and, as in **7.42**, $\qquad CD = \sqrt{(a^2 \tan^2 \theta + b^2 \sec^2 \theta)},$

therefore the required area $= 4ab$.

**7.44** It follows from **7.43** that if the tangent at $P$ meets the asymptotes in $L$, $L'$ then $LP = PL'$, since $LP = DC$ and $PL' = CD'$.

Also the triangle $LCL'$ is one quarter of the parallelogram $LL'K'K$, therefore the area of the triangle formed by the asymptotes and any tangent to the curve is $ab$.

**7.45** Geometrical properties of a hyperbola. A hyperbola has properties analogous to those of an ellipse proved in 6.6 and they can be established by similar proofs.

It will be noticed that the tangent to a hyperbola bisects the angle between the focal distances.

**7.5    The rectangular hyperbola.** When the asymptotes of a hyperbola are at right angles it is called **a rectangular hyperbola**.

The lines        $\dfrac{x}{a} - \dfrac{y}{b} = 0$   and   $\dfrac{x}{a} + \dfrac{y}{b} = 0$

are at right angles if $\dfrac{1}{a^2} - \dfrac{1}{b^2} = 0$, or if $a = b$; hence the equation of a rectangular hyperbola is

$$x^2 - y^2 = a^2 \qquad\qquad \dots\dots(1)$$

and its asymptotes are the lines

$$x - y = 0 \quad\text{and}\quad x + y = 0 \qquad\qquad \dots\dots(2).$$

**7.51    Equation of a rectangular hyperbola referred to its asymptotes as axes.** Let $x$, $y$ be the co-ordinates of a point $P$ on the rectangular hyperbola    $x^2 - y^2 = a^2$    $\dots\dots(1)$.

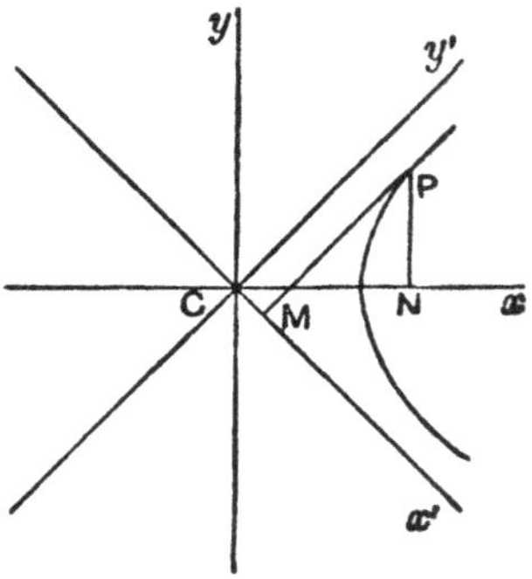

Let $x'$, $y'$ be the co-ordinates of $P$ referred to the asymptotes as axes.

In the figure

$$CN = x, \quad NP = y,$$
$$CM = x', \quad MP = y'.$$

Since the angle $x'Cx = \tfrac{1}{4}\pi$, and $CN$ is the projection on $Cx$ of $CM$ and $MP$, therefore

$$CN = CM \cos \tfrac{1}{4}\pi + MP \cos \tfrac{1}{4}\pi,$$

or        $x = (x' + y')\dfrac{1}{\sqrt{2}}$ .

Similarly $NP$ is the projection on $Cy$ of $MP$ minus $CM$, and

$$y = (y' - x')\dfrac{1}{\sqrt{2}}.$$

Substituting these values in (1), we get

$$\tfrac{1}{2}(x' + y')^2 - \tfrac{1}{2}(y' - x')^2 = a^2,$$

or        $x'y' = \tfrac{1}{2}a^2$        $\dots\dots(2).$

Similarly the equation of the conjugate hyperbola referred to the axes $Ox$, $Oy$ is        $y^2 - x^2 = a^2$        $\dots\dots(8)$

and, referred to the asymptotes as axes, it is

$$x'y' = -\tfrac{1}{2}a^2.$$

**7.52 Parametric representation of a rectangular hyperbola.** In 7.51 by transforming the axes of co-ordinates we showed that the equation of a rectangular hyperbola referred to its asymptotes as axes is

$$xy = \text{constant.}$$

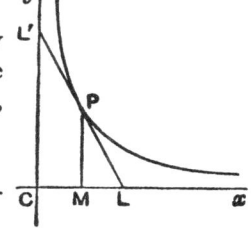

We may also deduce this result directly from 7.44 where it is proved that if the tangent at $P$ meets the asymptotes at $L$, $L'$ then $LP = PL'$ and the area

$$LCL' = \text{constant.}$$

For if $P$ is the point $(x, y)$ and $PM$ is perpendicular to the asymptote $Cx$, then

$$x = CM = \tfrac{1}{2}CL$$

and

$$y = MP = \tfrac{1}{2}CL';$$

but

$$CL \cdot CL' = \text{constant,}$$

therefore

$$xy = \text{constant.}$$

Now taking

$$xy = c^2 \qquad \qquad \ldots\ldots(1)$$

to be the equation of the curve, we see that it is satisfied by

$$x = ct, \quad y = \frac{c}{t} \qquad \qquad \ldots\ldots(2)$$

where $t$ is a variable parameter, so that for different values of $t$ the formulae (2) give the co-ordinates of all points on the curve.

As $t$ increases from 0 to $\infty$, $x$ is at first zero and $y$ infinite; for $t = 1$, we have $x = y = c$. This point is on the bisector of the angle between the asymptotes—i.e. a point on the transverse axis of the hyperbola. Then as $t$ continues to increase, $x$ increases without limit and $y$ decreases to zero.

Negative values of $t$ give the other branch of the hyperbola which lies in the third quadrant.

**7.53 Chord joining the points $t_1$, $t_2$.** The equation of the line joining the points $\left(ct_1, \frac{c}{t_1}\right)$, $\left(ct_2, \frac{c}{t_2}\right)$ is

$$\frac{x - ct_1}{c\,(t_1 - t_2)} = \frac{y - c/t_1}{c\left(\dfrac{1}{t_1} - \dfrac{1}{t_2}\right)},$$

which reduces to

$$x + t_1 t_2\, y = c\,(t_1 + t_2) \qquad \qquad \ldots\ldots(1).$$

From this we can deduce the equation of the tangent at the point $\left(ct, \dfrac{c}{t}\right)$, by putting $t_1 = t_2 = t$. This gives

$$x + t^2 y = 2ct \qquad \ldots\ldots(2).$$

Alternatively we may find the equation of the tangent by the method of 3.52.

**7.54 Normal at the point $ct$, $c/t$.** Since the normal is perpendicular to the tangent, therefore from 7.53 (2) its equation is

$$t^2 (x - ct) - \left(y - \frac{c}{t}\right) = 0,$$

or

$$t^3 x - ty = c\,(t^4 - 1).$$

**7.55 Polar of the point $(x', y')$.** It was shown in 3.22 (iii) that the equation of the tangent at the point $(x', y')$ on the curve $xy = c^2$ is

$$xy' + yx' = 2c^2.$$

It is easy to verify that this agrees with equation (2) of 7.53.

*Pole* and *polar* are defined in 4.51, and as in 4.51 or 5.72 it can be shown that the polar of the point $(x', y')$ with regard to the rectangular hyperbola $xy = c^2$ is

$$xy' + yx' = 2c^2.$$

**7.56 Examples.** The ellipse and the hyperbola possess so many analogous properties that a large number of theorems proved for the one can be proved in a similar way for the other; consequently our examples on the hyperbola will be confined for the most part to questions relating to the asymptotes and to rectangular hyperbolas.

*Prove that, if PQ be a chord of a rectangular hyperbola, normal at P, then the circle on PQ as diameter cuts the curve at the other extremity of the diameter through P.*

Let $P$ be the point $(ct, c/t)$ on the hyperbola $xy = c^2$; then the equation of the normal at $P$ is

$$t^3 x - ty = c\,(t^4 - 1) \quad (7.54)$$

and this meets the curve $xy = c^2$, where

$$t^3 x - \frac{c^2 t}{x} = c\,(t^4 - 1),$$

or

$$t^3 x^2 - (t^4 - 1)\,cx - c^2 t = 0,$$

i.e.

$$(x - ct)\,(t^3 x + c) = 0.$$

The root $ct$ is the abscissa of $P$, therefore the root $-c/t^3$ is the abscissa of $Q$, so that $Q$ is the point $x = -c/t^3$, $y = -ct^3$.

Then, by 4.21, the equation of the circle on $PQ$ as diameter is

$$(x - ct)\left(x + \frac{c}{t^3}\right) + \left(y - \frac{c}{t}\right)(y + ct^3) = 0.$$

The other extremity of the diameter through $P$ of the hyperbola is the point $(-ct, -c/t)$ and it is easily verified that the circle passes through this point.

### 7.6 Examples.

1. Show that the eccentricity of a rectangular hyperbola is $\sqrt{2}$.

2. Prove that the locus of the centre of a circle which touches two given non-concentric circles of unequal radii is an ellipse or a hyperbola.

3. $ABC$ is a triangle in which $C$ is a right angle. On $CA$, $CB$ points $P$, $Q$ are taken such that $CP \cdot CQ = CA \cdot CB$. Prove that the locus of the centroid of the triangle $CPQ$ is a rectangular hyperbola.

4. Show that the tangents to a rectangular hyperbola at the extremities of its latera recta pass through the vertices of the conjugate hyperbola.

5. $P$ is a point on a rectangular hyperbola whose centre is $C$ and a line is drawn through $C$ perpendicular to $CP$. Through $Q$, any point on the curve, lines are drawn parallel to the asymptotes meeting this line in $L$, $M$, show that $LPM$ is a right angle.

6. Prove that the intercept on any tangent to a hyperbola, made by its asymptotes, subtends a constant angle at either focus.

7. Prove that the line $lx + my = n$ touches the rectangular hyperbola $xy = c^2$, if
$$n^2 = 4lmc^2.$$

8. A tangent to a hyperbola of foci $S$, $S'$ meets the asymptotes in $L$, $L'$. Prove that the points $S$, $L$, $S'$, $L'$ are concyclic.

9. The tangents at two points $P$, $P'$ on a rectangular hyperbola meet an asymptote in $L$, $L'$ and $PP'$ meets it in $K$. Prove that
$$LK = KL'.$$

10. Prove that conjugate diameters of a rectangular hyperbola are equally inclined to the asymptotes.

11. Prove that the polars of a point with regard to two conjugate hyperbolas are parallel and equidistant from the centre.

**12.** Prove that in a rectangular hyperbola any chord $PP'$ subtends at the ends of a diameter $AA'$ angles which are either equal or supplementary.

**13.** Prove that if $CP$, $CD$ are conjugate radii of a hyperbola the orthocentre of the triangle $CPD$ lies on the line $ax = by$.

**14.** Prove that, if the tangent at $P$ to a rectangular hyperbola meets the asymptotes in $L$, $L'$ and the normal at $P$ meets the transverse axis in $G$, then $LGL'$ is a right angle.

**15.** Prove that if an ellipse and a hyperbola have the same foci they cut one another at right angles.

**16.** The perpendiculars from a point $P$ to the axes meet them in $M$, $N$ and the perpendicular bisector of $MN$ passes through a fixed point $C$ on one of the axes. Prove that the locus of $P$ is a rectangular hyperbola with centre at $C$.

**17.** Show that the locus of the middle point of a chord of the rectangular hyperbola $xy = c^2$ of constant length $2a$ is

$$(xy - c^2)(x^2 + y^2) = a^2 xy.$$

**18.** Prove that, if the position of a point on a rectangular hyperbola is determined by the variable $\theta$ where $x = c \tan \theta$, $y = c \cot \theta$, the locus of the intersection of tangents at the points $\theta$, $\theta + \alpha$ is

$$4(c^2 - xy) = (x + y)^2 \tan^2 \alpha,$$

$\alpha$ being a constant angle.

**19.** Tangents at right angles are drawn to a rectangular hyperbola and its conjugate. Show that they cut either asymptote in two points $K$, $K'$ such that the rectangle $CK \cdot CK'$ is equal to twice the square on the semi-axis, where $C$ is the centre.

**20.** Prove that the line joining the feet of the perpendiculars drawn to a pair of conjugate diameters of a rectangular hyperbola from any point $P$ on the hyperbola is parallel to the normal at $P$.

**21.** The normal at $P_1$ on the hyperbola $xy - c^2 = 0$ meets the curve again at $P_2$, the normal at $P_2$ meets the curve again at $P_3$ and so on. Prove that if $y_1$, $y_2$, $y_3$, ... $y_{n+1}$ are the ordinates of these points respectively $y_1^3 y_2 = y_2^3 y_3 = \cdots = y_n^3 y_{n+1} = -c^4.$

**22.** If $PN$ be the ordinate and $PG$ the normal of a point $P$ of a hyperbola, whose centre is $C$, and the tangent at $P$ intersect the asymptotes at $L$ and $L'$, show that half the sum of $CL$ and $CL'$ is the mean proportional between $CN$ and $CG$.

**23.** At the point of intersection of the rectangular hyperbola $xy = k^2$, and of the parabola $y^2 = 4ax$, the tangents to the hyperbola and parabola make angles $\theta$ and $\phi$ respectively with the axis of $x$.

Prove that $\qquad \tan \theta = -2 \tan \phi.$

**24.** Prove that, if $A$, $B$, $C$ are three points on a rectangular hyperbola, the curve passes through the orthocentre of the triangle $ABC$.

**25.** Prove that, in a rectangular hyperbola, the product of the focal distances of a point is equal to the square of the distance of the point from the centre.

**26.** Prove that, if $(c \tan \theta,\ c \cot \theta)$ and $(c \tan\theta',\ c \cot \theta')$ are two points on the hyperbola $xy = c^2$, and $\theta + \theta'$ is constant, then the chord joining the points passes through a fixed point on the conjugate axis of the hyperbola.

**27.** Show that the point whose co-ordinates are

$$\tfrac{1}{2}a\left(t+\frac{1}{t}\right), \quad \tfrac{1}{2}b\left(t-\frac{1}{t}\right)$$

lies on the hyperbola $x^2/a^2 - y^2/b^2 = 1$.

Prove that, if $C$ is the centre of the hyperbola and $S$ is either focus and if the tangent at the above point meets the asymptote $x/a = y/b$ at $X$ and meets the asymptote $x/a = -y/b$ at $Y$, then

$$t = CX/CS = CS/CY.$$

**28.** From any point on the normal at a given point $A$ on a rectangular hyperbola the other three normals to the curve are drawn. Show that the centroid of the feet of these three normals lies on the diameter of the hyperbola parallel to the normal at $A$.

**29.** A circle is drawn passing through any point $P$ on the hyperbola $x^2/a^2 - y^2/b^2 = 1$ and through the ends $A$, $A'$ of the transverse axis. The ordinate $NP$ is produced to meet the circle again in $Q$. Prove that, as the position of $P$ on the hyperbola varies, the locus of $Q$ is the hyperbola $x^2/a^2 - y^2/b'^2 = 1$, where $b' = a^2/b$.

# CHAPTER VIII

# POLAR CO-ORDINATES

**8.1 Polar co-ordinates.** The position of a point $P$ in a plane is determined when its distance $r$ from a fixed point $O$ and the angle $\theta$ that $OP$ makes with a fixed direction $OX$ in the plane are given.

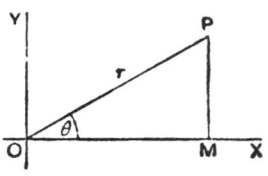

$r$ and $\theta$ are the polar co-ordinates of $P$ referred to $O$ as pole and $OX$ as initial line. $r$ is called the **radius vector** and $\theta$ is called the **vectorial angle** of the point $P$.

If $PM$ be perpendicular to $OX$, the Cartesian co-ordinates $x$, $y$ of $P$ are connected with the polar co-ordinates by the relations

$$x = OM = r \cos \theta, \quad \text{and} \quad y = MP = r \sin \theta \quad......(1);$$

and, conversely,

$$r^2 = x^2 + y^2 \quad \text{and} \quad \theta = \tan^{-1}(y/x) \quad ......(2).$$

These relations enable us to convert the equation of a locus from Cartesian to polar co-ordinates, or vice versa.

**8.12 Equation of a straight line.** Let $P$ be any point $(r, \theta)$ on the line, i.e. $OP = r$ and $POX = \theta$.

Let $M$ be the foot of the perpendicular from $O$ to the line, and let $M$ be the point $(p, \alpha)$, i.e.

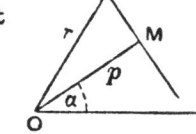

$$OM = p \quad \text{and} \quad MOX = \alpha.$$

Then since $OM = OP \cos MOP$,

therefore $p = r \cos(\theta - \alpha)$.

It follows that if we regard the line as determined by given constants $p$ and $\alpha$, its equation is

$$r \cos(\theta - \alpha) = p \quad ......(1).$$

If we expand the cosine and put $x$, $y$ for $r \cos \theta$, $r \sin \theta$ we have

$$x \cos \alpha + y \sin \alpha = p \quad ......(2),$$

as in **2.4**: and, conversely, (1) can be deduced from (2).

**8.13**   Since the equations of all straight lines are of the form

$$ax + by + c = 0 \qquad \ldots\ldots(3)$$

and can be converted into the form (2) by dividing by $\sqrt{(a^2 + b^2)}$, therefore the equation of every straight line in polar co-ordinates is of the form (1).

Further, the equation

$$\frac{A}{r} = B \cos \theta + C \sin \theta \qquad \ldots\ldots(4)$$

can be recognised at once as the equation of a straight line, because, on multiplying by $r$, it transforms into $A = Bx + Cy$, which is of the first degree.

But $\frac{A}{r} = B \cos \theta + C \sin \theta + D$ is *not* the equation of a straight line,

because on multiplying by $r$ it transforms into

$$A = Bx + Cy + D\sqrt{(x^2 + y^2)},$$

which is not of the first degree.

**8.14**   The equation of a **straight line through the pole $O$ is simply**

$$\theta = \alpha,$$

where $\alpha$ is the inclination of the line to $OX$.

The equation expresses an obvious property of the straight line. It also follows from the Cartesian form of equation $y = x \tan \theta$, for this implies that $\theta$ is constant.

**8.2   Equation of a circle.**

(i) When the centre is at the pole and $a$ is the radius, the equation is

$$r = a \qquad \ldots\ldots(1).$$

(ii) When the pole is on the circumference and $a$ is the diameter.

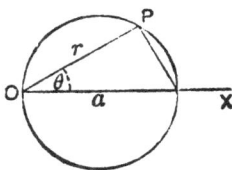
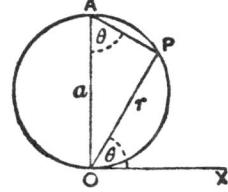

If we take the diameter through $O$ as initial line, the equation is seen to be

$$r = a \cos \theta \qquad \ldots\ldots(2).$$

If we take the tangent at $O$ as initial line, the equation is seen to be

$$r = a \sin \theta \qquad \ldots\ldots(3).$$

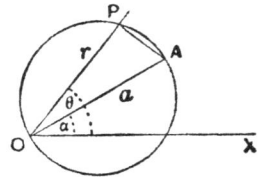

If the diameter $OA$ makes an angle $\alpha$ with $OX$, the equation is seen to be

$$r = a \cos (\theta - \alpha) \quad \ldots\ldots(4).$$

(iii) When the centre is the point $(a, \alpha)$ and $c$ is the radius.

Let $C$ be the centre, and $P$ the point $(r, \theta)$ on the circle.

Then

$$CP^2 = OP^2 + OC^2 - 2OP.OC \cos POC,$$

therefore

$$c^2 = r^2 + a^2 - 2ra \cos (\theta - \alpha) \ldots\ldots(5)$$

and this is the equation required.

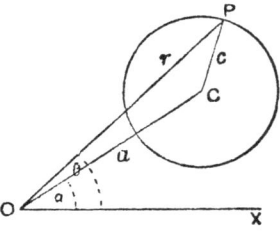

**8.21** We note that the equations (1)–(5) of **8.2** are easily converted into Cartesian co-ordinates. Thus on multiplying (3) by $r$, we have

$$r^2 = ar \sin \theta, \quad \text{or} \quad x^2 + y^2 = ay.$$

The transformation of the other equations is left as an exercise for the student.

**8.3 Conics referred to their centres as pole.** The equations of ellipses and hyperbolas are of the form

$$\frac{x^2}{a^2} \pm \frac{y^2}{b^2} = 1,$$

and by substituting $x = r \cos \theta$, $y = r \sin \theta$ the polar equations can be written in the usual form

$$\frac{1}{r^2} = \frac{\cos^2 \theta}{a^2} \pm \frac{\sin^2 \theta}{b^2}.$$

**8.4 Conics referred to a focus as pole.** Let $S$ be the focus, and let $SX$ perpendicular to the directrix $XM$ meet the curve at the vertex $A$. Let $P$ be the point $(r, \theta)$ on the curve, so that $SP = r$ and $PSA = \theta$.

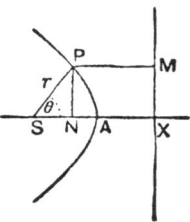

Then if $e$ is the eccentricity and $l$ the semi-latus rectum, and $PM$, $PN$ are perpendicular to the directrix and the axis, we have

$$SP = ePM = eNX = e\,(SX - SN)$$
$$= l - eSP \cos PSA,$$

or $\qquad r = l - er \cos \theta.$

Whence we get $\qquad \dfrac{l}{r} = 1 + e \cos \theta$ $\qquad$ ......(1),

which is the standard equation of a conic referred to a focus as pole, $\theta$ being measured from the line through $S$ perpendicular to the corresponding directrix.

The equation of the directrix is easily seen to be

$$r \cos \theta = SX = l/e,$$

or $\qquad \dfrac{l}{r} = e \cos \theta$ $\qquad$ ......(2).

**8.5 The formula $\tan \phi = r \dfrac{d\theta}{dr}$.** Let $P$, $P'$ be any two neighbouring points on a plane curve.

Let $P$ be the point $(r,\ \theta)$ and $P'$ be $(r + \delta r,\ \theta + \delta\theta)$, so that the angle $POP'$ is $\delta\theta$. Then if $P'PR$ is the chord joining the points, the angle

$$OP'R = OPR - \delta\theta,$$

and, from the triangle $OPP'$, we have

$$OP' \sin OP'R = OP \sin OPR,$$

or $\qquad (r + \delta r) \sin (OPR - \delta\theta) = r \sin OPR$ $\qquad$ ......(1).

Now for a small angle we may put, as a first approximation, $\sin \delta\theta = \delta\theta$ and $\cos \delta\theta = 1$, so that (1) is equivalent to

$$(r + \delta r)\,(\sin OPR - \delta\theta \cos OPR) = r \sin OPR,$$

or $\qquad \delta r . \sin OPR - r\,\delta\theta . \cos OPR - \delta r\,\delta\theta . \cos OPR = 0,$

whence $\qquad \tan OPR = r \dfrac{\delta\theta}{\delta r} - \delta\theta$ $\qquad$ ......(2).

Now let $P'$ move up to $P$, then $\delta\theta$ and $\delta r$ both tend to zero and $\delta\theta/\delta r$ tends to the derivative $d\theta/dr$.

Also $P'P$ becomes, in the limit, the tangent $PT$ to the curve at $P$, and $OPR$ becomes the angle $\phi$ between the tangent and the radius vector, and the formula (2) becomes

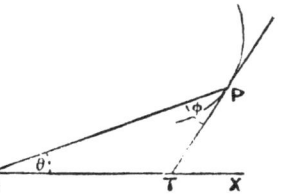

$$\tan\phi = r\,\frac{d\theta}{dr} \qquad \ldots\ldots(3).$$

**8.6 Curve tracing in polar co-ordinates.** The formula of the last article can be used to establish the equation in polar co-ordinates of the tangent at any point on a curve, but the geometrical properties of curves are generally much more easily investigated in Cartesian co-ordinates, and a more practical use of formula (3) is in the tracing of curves from their polar equations.

*For example, consider the curve $r = a\,(1 + \cos\theta)$.*

The equation remains unaltered when the sign of $\theta$ is changed, so that the curve must be symmetrical about the initial line.

Then by differentiation $\dfrac{dr}{d\theta} = -a\sin\theta,$

so that $\tan\phi = r\,\dfrac{d\theta}{dr} = -\dfrac{1+\cos\theta}{\sin\theta} = -\cot\tfrac{1}{2}\theta$

$$= \tan\tfrac{1}{2}\,(\pi+\theta),$$

and therefore $\phi = \tfrac{1}{2}\,(\pi+\theta).$

Now when $\theta = 0$, then $r = 2a$ and $\phi = \tfrac{1}{2}\pi$; i.e. the curve is at right angles to the initial line.

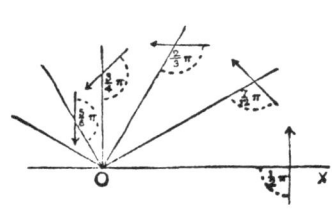
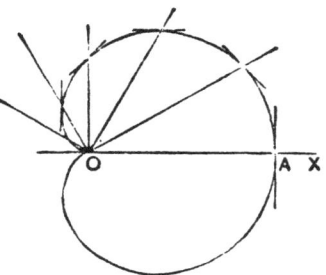

Then as $\theta$ increases from 0 to $\tfrac{1}{2}\pi$, $r$ decreases steadily from $2a$ to $a$, and $\phi$ increases steadily from $\tfrac{1}{2}\pi$ to $\tfrac{3}{4}\pi$.

Finally as $\theta$ increases from $\frac{1}{2}\pi$ to $\pi$, $r$ decreases steadily from $a$ to 0, and $\phi$ increases steadily from $\frac{3}{4}\pi$ to $\pi$.

Values may also be plotted for intermediate values of $\theta$ such as $\frac{1}{4}\pi$ and $\frac{1}{3}\pi$. The directions of the tangents are shown at five points on the curve in the first figure and the curve is then easily drawn as in the second figure.

This curve is called the *Cardioid*. It is equally well represented by the equation $r = a\,(1 - \cos\theta)$, as this form of the equation only interchanges the relative positions of the pole $O$ and the vertex $A$.

**8.7 Perpendicular from pole to tangent.** If $p$ is the length of the perpendicular $OY$ from the pole $O$ to the tangent $PT$ at a point $(r, \theta)$ on a curve, we have

$$p = OY = OP \sin\phi = r \sin\phi.$$

But $\quad \tan\phi = r\dfrac{d\theta}{dr} \quad$ **(8.5)**,

therefore

$$\frac{1}{p^2} = \frac{\operatorname{cosec}^2\phi}{r^2} = \frac{1}{r^2} + \frac{\cot^2\phi}{r^2}$$

or
$$\frac{1}{p^2} = \frac{1}{r^2} + \frac{1}{r^4}\left(\frac{dr}{d\theta}\right)^2 \qquad \ldots\ldots(1).$$

Thus $p$ can be determined when the polar equation of the curve is known.

**8.71 Pedal, or $p$, $r$, equations.** The equation of a curve can be expressed as a relation between the radius vector $r$ and the distance $p$ of the tangent from the pole, and such an equation is known as a *pedal*, or $p$, $r$, equation.

Starting from the equation of the curve in polar co-ordinates, say
$$r = f(\theta) \qquad \ldots\ldots(1),$$

then, from 8.7 (1), we have
$$\frac{1}{p^2} = \frac{1}{r^2} + \frac{1}{r^4}\left(\frac{dr}{d\theta}\right)^2 \qquad \ldots\ldots(2),$$

and the elimination of $\theta$ between (1) and (2) will give the required relation between $p$ and $r$.

For example, for the cardioid of **8.6** we have

$$r = a (1 + \cos \theta) \qquad \ldots\ldots(8)$$

and

$$\frac{dr}{d\theta} = - a \sin \theta,$$

so that

$$\frac{1}{p^2} = \frac{1}{r^2} \left\{ 1 + \frac{\sin^2 \theta}{(1 + \cos \theta)^2} \right\}$$

$$= \frac{1}{r^2} \{1 + \tan^2 \tfrac{1}{2}\theta\}$$

$$= \frac{1}{r^2} \sec^2 \tfrac{1}{2}\theta.$$

Therefore $\qquad p^2 = r^2 \cos^2 \tfrac{1}{2}\theta = \tfrac{1}{2}r^2 (1 + \cos \theta) \qquad \ldots\ldots(4)$,

so that by eliminating $\theta$ between (3) and (4) we get

$$2ap^2 = r^3$$

as the pedal equation of the cardioid.

**8.72  Pedal equations.** Pedal equations are specially useful in certain dynamical problems and, for well known curves such as the circle and conic sections, they can be found by simple geometrical methods.

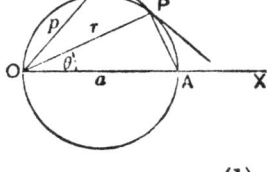

(i) *Circle $r = a \cos \theta$.*

Since in the figure the angles $OPY$ and $OAP$ are equal, therefore by similar triangles $\qquad OY : OP = OP : OA$,

or $\qquad\qquad pa = r^2 \qquad\qquad \ldots\ldots(1)$

is the pedal equation.

(ii) *Parabola with focus as pole.*

It is shown in **5.4** that in the parabola $SY^2 = SP \cdot SA$, or

$$p^2 = ar \qquad\qquad \ldots\ldots(2),$$

which is therefore the pedal equation of the parabola.

(iii) *Ellipse with focus as pole.*

In the figure of **6.6**, if $SP = r$, $S'P = r'$, $SY = p$ and $S'Y' = p'$, we get from similar triangles $SPY$, $S'PY'$ as in **6.6** (8)

$$\frac{p}{r} = \frac{p'}{r'},$$

therefore each fraction also $\qquad = \sqrt{\left(\frac{pp'}{rr'}\right)}.$

But by 6.6 (7)    $pp' = SY . S'Y' = b^2$

and    $r' = 2a - r,$

so that    $\dfrac{p}{r} = \dfrac{b}{\sqrt{\{r\,(2a-r)\}}}$

or    $\dfrac{b^2}{p^2} = \dfrac{2a}{r} - 1$    ......(3)

and this is the pedal equation of the ellipse.

In like manner the pedal equation of the hyperbola with focus as pole is

$$\frac{b^2}{p^2} = \frac{2a}{r} + 1$$    ......(4),

the only difference in the proof being that by 7.11 $r' = 2a + r.$

(iv) *Ellipse with centre as pole.*

From 6.53 and 6.54 we have

$$CP^2 + CD^2 = a^2 + b^2$$

and    $$CD . PF = ab.$$

Therefore, putting $CP = r$ and $PF = p$, and eliminating $CD$, we get

$$\frac{a^2 b^2}{p^2} = a^2 + b^2 - r^2$$    ......(5),

which is the required pedal equation.

**8.8  Special curves.**

(i) *The lemniscate.*    $r^2 = a^2 \cos 2\theta.$

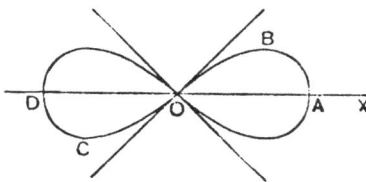

The equation remains unaltered if the sign of $\theta$ is changed, so that there is symmetry about the initial line. Also, for real values of $r$, $\cos 2\theta$ must be positive, therefore values of $2\theta$ between $\tfrac{1}{2}\pi$ and $\tfrac{3}{2}\pi$, or values of $\theta$ between $\tfrac{1}{4}\pi$ and $\tfrac{3}{4}\pi$ do not give real values for $r$. Further there are two real values of $r$ equal but opposite in sign, namely $\pm a\sqrt{(\cos 2\theta)}$, for all values of $\theta$ between $-\tfrac{1}{4}\pi$ and $\tfrac{1}{4}\pi$.

Again by differentiating we find that

$$r\frac{dr}{d\theta} = -2a^2 \sin 2\theta;$$

therefore    $\tan \phi = r\dfrac{d\theta}{dr} = -\tfrac{1}{2}\cot 2\theta.$

Hence when $\theta = 0$, $r = \pm a$ and $\tan \phi$ is infinite so that $\phi = \frac{1}{2}\pi$ and the curve is at right angles to the initial line. Then, as $\theta$ increases from 0 to $\frac{1}{4}\pi$, $r$ decreases steadily to zero, and $\tan \phi$ is negative for all the intermediate values and becomes zero when $\theta = \frac{1}{4}\pi$, so that $\phi$ increases from $\frac{1}{2}\pi$ to $\pi$, and as the curve approaches $O$ the tangent tends to coincide in direction with the radius vector.

Thus as $\theta$ increases from 0 to $\frac{1}{4}\pi$, the positive values of $r$ give a branch $ABO$ and the negative values a similar branch $DCO$; and for the values of $\theta$ between $\frac{3}{4}\pi$ and $\pi$ we get the remaining branches completing a figure of eight.

(ii) *The equiangular spiral.* $r = ae^{\theta \cot a}$.

It is evident from the equation that, if $\alpha$ is an acute angle so that $\theta \cot \alpha$ is positive, then as $\theta$ increases so does $r$ and the curve is a spiral with an ever increasing radius.

Further, since $\qquad \log r = \log a + \theta \cot \alpha$,

therefore by differentiation $\dfrac{1}{r}\dfrac{dr}{d\theta} = \cot \alpha$.

Hence

$$\tan \phi = r \frac{d\theta}{dr} = \tan \alpha, \quad \text{or} \quad \phi = \alpha.$$

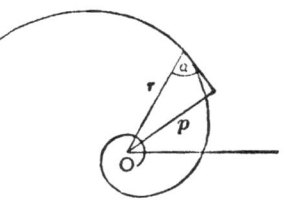

Therefore the curve possesses the property that the angle between the tangent and the radius vector is a constant angle, and hence the name of the curve.

It follows that the pedal equation of the curve is $p = r \sin \alpha$.

### 8.9 Examples.

1. Find the polar equations of the curves whose Cartesian equations are:

(i) $x^2 + y^2 + 2gx + 2fy + c = 0$;

(ii) $x^2 + y^2 = ax$;

(iii) $x^2 - y^2 = c^2$;

(iv) $xy = \frac{1}{2}c^2$;

(v) $x^3 - 3xy^2 = a^3$.

2. Find the Cartesian equations corresponding to the following equations in polar co-ordinates:

(i) $r \sin \theta + a = 0$;      (ii) $r = a \sin \theta + b \cos \theta$;

(iii) $r^2 \cos 2\theta = a^2$;      (iv) $\dfrac{l}{r} = 1 + e \cos \theta$;

(v) $r^2 \sin 3\theta = a^3$;      (vi) $r^2 = a^2 + b^2 \cos 2\theta$.

**3.** Using the polar equation of a conic with focus as pole, show that the semi-latus rectum is a harmonic mean between the segments of a focal chord.

**4.** Draw the curve $r = a \cos 2\theta$, noting carefully when $r$ changes sign and showing that the curve forms a double figure of eight.

**5.** Draw the curve $r = a (2 + \cos \theta)$.

**6.** Draw the curve $r = a (1 + 2 \cos \theta)$, showing that it consists of two loops one within the other.

**7.** Deduce from the polar equations of the curves the pedal equations obtained in 8.72.

**8.** Find the pedal equation of:

     (i) the rectangular hyperbola $xy = c^2$;

     (ii) the lemniscate $r^2 = a^2 \cos 2\theta$;

     (iii) the curve $r^n = a^n \cos n\theta$;

     (iv) the spiral $r\theta = a$.

**9.** Prove that, for the curve $r\theta = a$, if a perpendicular through the pole be drawn to the radius vector at any point, the length intercepted on it by the tangent is constant.

**10.** Prove that, for the curve $r = a \cos (\theta - \alpha)$, $\phi = \frac{1}{2} \pi + \theta - \alpha$.

**11.** Show that, for the curve $r = a (1 + \cos \theta)$, the locus of the foot of the perpendicular from the pole to the tangent is $r = 2a \cos^3 \frac{1}{3}\theta$.

**12.** Prove that, for the curve $r = a \sec^3 \frac{1}{3}\theta$, the locus of the foot of the perpendicular from the pole to the tangent is a parabola.

**13.** Trace the curve $r^2 = a^2 \cos \theta$.

**14.** Prove that the pedal equation of the curve $r \cos n\theta = a$ is

$$\frac{1}{p^2} = \frac{1 - n^2}{r^2} + \frac{n^2}{a^2}.$$

# EASY EXAMPLES

## 9.1 The Straight Line.

1. Find the distance between the points $A$, $B$ whose co-ordinates are $(3, 7)$ and $(11, 13)$. Also find the co-ordinates of the point which divides $AB$ in the ratio $3 : 1$.

2. Find the equations of the sides of a triangle whose corners are at the points $(3, 4)$, $(-3, 2)$, $(2, -4)$.

3. In the last Example find the equations of the perpendiculars from the corners of the triangle to the opposite sides and verify that they meet in a point.

4. Find the distance between the points $A$ $(2, 3)$ and $B$ $(14, 8)$. Find also the equation of the line $AB$ and the distance from it of the point $C$ $(5, 6)$. What is the area of the triangle $ABC$?

5. The equations of the sides of a triangle are $x + y = 5$, $2x - y + 4 = 0$ and $x - 4y + 4 = 0$. Find the tangents of the angles of the triangle.

6. Find the equation of a line parallel to the line $5x + 4y = 9$ and making an intercept $-5$ on the $x$-axis.

7. Find the equations of two straight lines which pass through the intersection of the lines $3x + 4y = 7$, $5x - 2y = 3$ and are parallel and perpendicular respectively to the line $2x + 7y = 3$.

8. Find the value of $k$ if the three lines $2x + 5y = 12$, $7x + 2y = 11$ and $kx - 3y = 10$ meet in a point.

9. Prove that the points $(x', y')$, $(x'', y'')$ and

$$\left( \frac{x' + kx''}{1 + k}, \frac{y' + ky''}{1 + k} \right)$$

are collinear.

10. Show that the points $(2, 3)$ and $(-1, 2)$ are on opposite sides of the line $2x - 8y + 7 = 0$. What are their distances from the line?

11. Two straight lines $AB$, $CD$ bisect one another at right angles. The co-ordinates of $A$, $B$ and $C$ are $(1, 5)$, $(3, 1)$ and $(-1, 1\frac{1}{2})$. Find the co-ordinates of $D$ and the area $ABCD$.

12. Prove that the line $2x + 3y = 5$ divides the join of the points $(8, -5)$, $(2, 1)$ in the ratio $7 : 1$.

**13.** Find the equations of the two straight lines which pass through the point (3, 2) and make angles of 45° with the line $4x - 5y = 6$.

**14.** Find the equation of the perpendicular bisector of the join of the points (3, 5), ($-6$, 7).

**15.** From a point $A$ (2, 5) a perpendicular $AB$ is drawn to the line $3x - 4y = 8$. $AB$ is then produced to $C$ so that $BC = 3AB$. Find the co-ordinates of $C$.

**16.** The sides $BC$, $CA$, $AB$ of a triangle have equations $3x + 4y = 9$, $4x - 3y + 6 = 0$, $5x - 12y + 12 = 0$. Find the equations of the bisectors of the interior angles of the triangle and the co-ordinates of their point of concurrence.

**17.** $ABC$ and $DBC$ are two triangles such that $DB = AB$ and $DC = AC$. The co-ordinates of $A$, $B$, $C$ are (3, 1), (2, 2), ($-1$, 3). Find the co-ordinate of $D$. Also find the equations of $AD$ and $BC$ and verify that these lines are at right angles.

**18.** Find the centre of the inscribed circle of the triangle formed by the lines $3x - 4y + 8 = 0$, $4x + 3y - 9 = 0$, $y = 6$.

**19.** Find the equations of two lines passing through the intersection of the lines $x + 3y = 5$ and $4x - y + 2 = 0$, and perpendicular and parallel respectively to the line $5x - 3y + 3 = 0$.

**20.** Find the equation of the line which passes through the intersection of the lines $2x + 3y = 4$ and $3x - y + 2 = 0$ and also through the intersection of the lines $x + y = 0$, $4x + y - 3 = 0$.

**21.** Prove that the points $A$ (4, 2) and $C$ ($-3$, 1) subtend a right angle at $B$ ($-\frac{17}{13}$, $-\frac{20}{13}$). Find the co-ordinates of $D$, the remaining corner of the rectangle $ABCD$, and verify that $DB = AC$.

**22.** The points $A$ (4, 2), $B$ (1, $-2$), $C$ ($-3$, 1) and $D$ are the corners of a parallelogram. Find the co-ordinates of $D$ and prove that the parallelogram is equal in area to the square on $OD$ where $O$ is the origin.

**23.** Show that the equation $6x^2 + 7xy - 3y^2 + x + 7y - 2 = 0$ represents two straight lines, and find the tangents of the angles between them.

**24.** Find the separate equations of the two straight lines represented by the equation $x^2 - 2xy \operatorname{cosec} \theta + y^2 = 0$. Also find the angles between the lines, and show that these angles are bisected by the lines $x^2 - y^2 = 0$.

### 9.2  Curves and Loci.

1.  Find the equations of the tangents to the curves $y = x^3 - x$ and $y = x^2 - 3x + 2$ at their point of intersection, and the angle at which the curves cut.

2.  Prove that the line $3y = 9x - 16$ is a tangent to the curve $3y = x^3 - 3x$, and find the co-ordinates of the point of contact.

3.  Prove that the tangents to the curve $y = x^3 - 6x^2 + 13x - 6$ at the points $(1, 2)$ and $(3, 6)$ are parallel, and that the tangent at the point $(2, 4)$ is inclined to them at an angle $\tan^{-1}(3/5)$.

4.  Prove that the line $y = 5x$ is a tangent to the curve

$$y = x^3 - 2x^2 + x + 8,$$

and find the co-ordinates of the point of contact. In what other point does the line $y = 5x$ meet the curve?

5.  Find the equation of the tangent to the curve $y = 3x^2 - 2x - 1$ at the point $(2, 7)$. Also find the co-ordinates of the point on the curve the tangent at which is perpendicular to the tangent at $(2, 7)$.

6.  Find the equation of the normal to the curve $y = 2x^2 - 3x - 5$ at the point $(2, -3)$, and find the length of the subnormal at this point.

7.  Find the equation of the tangent to the curve

$$3y = x^3 - 6x^2 - 12x + 1$$

at the point $(-1, 2)$, and find the co-ordinates of another point on the curve where the tangent is parallel to that at $(-1, 2)$.

8.  Prove that at every point of the curve $y = x^3 - 3x^2 + 3x + 7$ the gradient is positive. Table the values of $y$ and $dy/dx$ at the points $x = -1, 0, 1, 2, 3$ and make a rough drawing of the curve.

9.  Find the equation of the tangent and normal to the curve $y = (x - 1)^4$ at the point $(2, 1)$. Find the area of the triangle which the tangent and normal make with the $x$-axis.

10.  Show that the curve $y = 1 - 2x + x^3$ cuts the $x$-axis in three real points. Find the gradient of the curve at the point $(1, 0)$; also find the points at which the gradient vanishes and make a rough drawing of the curve.

11.  Find the gradients of the curve $y = (x - 1)^2 (x - 2)$ at the points where it crosses the axes. Find also the points at which the gradient vanishes and make a rough drawing of the curve.

12. Find the points on the curve $y = x^3 - x^2 + 1$ at which the gradient is unity. Find the equations of the tangents and normals at these points.

13. Find the points in which the curve $y = 1 + 2x - 3x^2$ cuts the co-ordinate axes and find the gradients of the curve at these points. At what point does the gradient vanish? Make a rough drawing of the curve.

14. Show that the curve $y = (x^2 - 1)^2$ is symmetrical about the $y$-axis. Find its gradient and the points at which the gradient vanishes. In what points does the line $y = 9$ cut the curve? Make a rough drawing of the curve.

15. Show that the curve $y = x^4 - 4x^2$ is symmetrical about the $y$-axis. Find the gradients of the curve at its intersections with the $x$-axis. Find also the points where the gradient vanishes and make a rough drawing of the curve.

16. Find the gradient of the curve $y = x^2 - 3x - 4$, and the point on the curve at which the gradient vanishes. Also find the gradient at the points where the curve crosses the $x$-axis and make a rough drawing of the curve. To what point on the $x$-axis must the origin be moved so that the curve may become symmetrical about the $y$-axis, and what will be the equation of the curve when the position of the origin is so changed?

17. Show that the curve whose equation is $27y = x^3 - 27x$ is symmetrical in opposite quadrants. Prove that the tangent to the curve at the origin is $x + y = 0$. Find the gradient at the other intersections with the $x$-axis. Also find at what points the tangent is parallel to the $x$-axis, and make a rough drawing of the curve.

18. Find the equation of the locus of a point which moves so that its distance from the point $(2, 1)$ is twice its distance from the line $3x + 4y = 2$.

19. Find the locus of a point which moves so that its distance from the origin is half its distance from the line $x + 2y = 3$.

20. A point $P$ moves so that $PA^2 + PB^2 = AB^2$, where $A$, $B$ are the points $(2, 0)$, $(0, 3)$. Find the equation of the locus of $P$.

21. Find the locus of a point whose distance from the point $A$ $(3, 0)$ is three times its distance from the point $B$ $(-1, 0)$. Prove that the tangents to the locus at the points $(0, 0)$ and $(-3, 0)$ are parallel to the $y$-axis.

22. The co-ordinates of a point are given in terms of a variable parameter $t$ by the relations $x = t/(t+1)$, $y = t/(t-1)$. Find the equation of the locus of the point, and prove that the equation of the tangent at any point $t$ on the locus is $(t+1)^2 x + (t-1)^2 y = 2t^2$.

23. The co-ordinates of a point are given by the relations $x = t + \dfrac{1}{t}$, $y = t - \dfrac{1}{t}$, where $t$ is a variable parameter. Find the equation of the locus of the point for different values of $t$, and also the equations of the tangent at the point denoted by $t$.

### 9.3  The Circle.

1. Prove that the circle of centre $(3, 4)$ and radius 5 passes through the origin. Find the equations of the tangent at the origin and of the parallel tangent. In what other points does the circle cut the axes?

2. Find the equation of a circle having the points $(2, 1)$ and $(1, -3)$ for the ends of a diameter. Find the co-ordinates of the ends of the perpendicular diameter.

3. Find the equation of the tangent to the circle $x^2 + y^2 = 1$ at the point $(\frac{4}{5}, \frac{3}{5})$. Find the two points in which this tangent cuts the axes and also find the equations of the other tangents to the circle that can be drawn from the last two points.

4. For what values of $k$ does the line $y = kx$ touch the circle $x^2 + y^2 + 2x + 4y + 1 = 0$? Find the co-ordinates of the points of contact.

5. Show that the points $(5, 8)$ and $(-1, 0)$ are the opposite ends of a diameter of the circle $x^2 + y^2 - 4x - 8y - 5 = 0$. Find the equations of the tangents at these points.

6. Find the equation of the circle through the points $(3, 0)$, $(0, 4)$ and $(1, 2)$. Find the equation of the tangent at the point $(3, 13)$.

7. Find the equation of the circle which passes through the points $(3, 0)$, $(0, 4)$, $(0, 6)$. In what other point does the circle cut the $x$-axis?

8. Find where the line $x + y = 5$ cuts the circle

$$3x^2 + 3y^2 - 11x - 11y + 16 = 0?$$

What are the equations of the tangents parallel to the line?

**9.** Find the co-ordinates of the points in which the line $x + 2y = 7$ cuts the circle $x^2 + y^2 - 13x - 13y + 52 = 0$. Find the equations of the tangents at these points and the co-ordinates of the point in which these tangents intersect.

**10.** Prove that, if $(x', y')$ is the middle point of a chord of a circle $x^2 + y^2 = a^2$, then the equation of the chord is $xx' + yy' = x'^2 + y'^2$.

**11.** Find the co-ordinates of the middle point of the chord of the circle $x^2 + y^2 = 25$ which lies along the line $3x - 4y = 7$. Also find the points of contact of the parallel tangents.

**12.** Find the co-ordinates of the middle point of the chord of the circle $x^2 + y^2 - 4x + 6y + 1 = 0$ which lies along the line $2x - 3y = 12$.

**13.** Prove that the line $2x + y - 7 = 0$ is the equation of a diameter of the circle $x^2 + y^2 - 6x - 2y + 5 = 0$, and find the equations of the tangents at the ends of this diameter.

**14.** Find the angle which the circle $x^2 + y^2 - 6x - 4y + 9 = 0$ subtends at the origin.

**15.** Find the equations of the circles which pass through the point $(-2, 1)$ and touch both the co-ordinate axes.

**16.** Prove that the locus of points at which the circle

$$x^2 + y^2 - 4x - 6y + 4 = 0$$

subtends a right angle is the circle $x^2 + y^2 - 4x - 6y - 5 = 0$.

**17.** Find the equations of the tangents to the circle

$$x^2 + y^2 - 4x + 2y - 8 = 0$$

which are parallel to the line $2x + 3y = c$.

**18.** Find the equation of the chord of the circle

$$x^2 + y^2 - 4x + 2y - 4 = 0$$

which has $(3, 1)$ as its middle point.

**19.** Show that the points $(0, 0)$, $(2, 1)$, $(3, 3)$ and $(1, 2)$ are the corners of a rhombus, and find the equation of its inscribed circle.

**20.** Find the equation of the circle which has the same centre as the circle $x^2 + y^2 + 6x - 10y + 18 = 0$ and passes through the point $(6, 7)$. Show that the origin is outside the given circle and inside the other.

**21.** Find the locus of the centre of a circle

$$x^2 + y^2 + 2gx + 2fy + c = 0$$

which is such that the tangents drawn to the circle from the origin are always at right angles.

### 9.4  The Parabola.

1.  Find the value of $c$ if the parabola $y^2 = 4ax$ intercepts a length $4a$ on the line $y = x + c$.

2.  What is the equation of a parabola which is symmetrical about the $x$-axis, touches the $y$-axis at the origin and has a latus rectum of length 8? What are the co-ordinates of the focus? What are the co-ordinates of a point on the curve at a distance 20 from the focus?

3.  The parabola $y^2 = 4ax$ passes through the point $P$ (6, 6). If $S$ is the focus find the co-ordinates of the other point in which $PS$ meets the curve.

4.  Find the value of $a$ if the parabola $y^2 = 4ax$ touches the line $y = 3x + 4$. What are the co-ordinates of the point of contact?

5.  Prove that the line $ty = x + at^2$ touches the parabola $y^2 = 4ax$, and find the co-ordinates of the point of contact. Also find the co-ordinates of the foot of the perpendicular to this tangent from the focus of the parabola.

6.  Prove that, if $P$ is a point on a parabola $y^2 = 4ax$ whose vertex is $A$, and $PL$ at right angles to $AP$ meets the $x$-axis in $L$, and $PN$ is the ordinate of $P$, then $NL = 4a$.

7.  $PN$ is the ordinate at a point $P$ on a parabola and the normal at $P$ meets the axis in $G$. Prove that $PG$ is equal to the ordinate that passes through the middle point of $NG$.

8.  Find the locus of a point $P$ which moves so that its distance from the point (1, 0) is equal to its distance from the line $x + 1 = 0$. Also find the co-ordinates of the middle point of the chord of this locus which lies along the line $3y = 2x + 4$.

9.  The normal at $P$ to a parabola meets the axis in $G$ and $S$ is the focus. Prove that, if the triangle $SPG$ is equilateral, then $SP$ is equal to the latus rectum.

10.  $A$ is the vertex of a parabola $y^2 = 4ax$ and $LL'$ is the latus rectum. Prove that the diameter of the circle $LAL'$ is $5a$.

11.  The chord joining the points $(x_1, y_1)$, $(x_2, y_2)$ on the parabola $y^2 = 4ax$ cuts the axis at $C$. Prove that, if $A$ is the vertex, $x_1 x_2 = AC^2$.

12.  $C$ is a fixed point $(0, c)$ on the axis $Oy$ and $Q$ is a variable point on the line through $C$ parallel to $Ox$. A point $P$ is taken on $OQ$ so that the ordinate of $P$ is equal to $CQ$. Prove that the locus of $P$ is the parabola $y^2 = cx$.

**13.** The chord $PQ$ of a parabola passes through the focus. If $P$ is the point $(at^2, 2at)$ what are the co-ordinates of $Q$?

**14.** The chord $PQ$ of a parabola is the normal at $P$. If $P$ is the point $(at^2, 2at)$ what are the co-ordinates of $Q$?

**15.** Find the value of $a$ if the parabola $y^2 = 4ax$ touches the line $2y = x + 8$. Find the equation of the normal at the point of contact, and the co-ordinates of the second point of intersection of the normal and the curve.

**16.** Find the equations of the tangents to the parabola $y^2 = 16x$ at the points $(36, 24)$ and $(\frac{4}{9}, -\frac{8}{3})$ and verify that they intersect on the directrix.

**17.** Prove that the line $x - 6y + 36 = 0$ touches the parabola $y^2 = 4x$ and find the co-ordinates of the point of contact. Find also the co-ordinates of the foot of the perpendicular drawn to this tangent from the focus of the parabola.

**18.** Prove that the lines $2y = 6x + 1$ and $8y = 32x + 8$ both touch the parabola $y^2 = 6x$, and find the equation of the line joining the points of contact.

**19.** Find the value of $c$ if the parabola $y^2 = 8x$ intercept a length 8 on the line $y = x + c$.

**20.** Points on a parabola are represented parametrically by the relations $x = at^2$, $y = 2at$, where $t$ is the variable parameter. Normals are drawn at the points $t = 2$ and $t = 1$. Prove that they intersect on the curve.

**21.** Prove that, if the points $(x', y')$, $(x'', y'')$ are the ends of a chord passing through the focus of a parabola $y^2 = 4ax$, then $x'x'' = a^2$ and $y'y'' = -4a^2$.

**22.** From the focus $S$ of a parabola a line is drawn parallel to the tangent at $P$ $(at^2, 2at)$ meeting the line $y = 2at$ in $Q$. Prove that the locus of $Q$ is the parabola $y^2 = 2a(x - a)$.

**23.** Prove that tangents and normals to a parabola at the points $(at^2, 2at)$, $(a/t^2, -2a/t)$ enclose a rectangle of area $a^2(t + 1/t)^3$.

### 9.5 The Ellipse.

**1.** An ellipse has its foci at the points $(2, 0)$, $(-2, 0)$ and passes through the point $(2, 3)$. Find its equation.

**2.** Prove that the line $x + 2y = 8$ touches the ellipse $3x^2 + 4y^2 = 48$, and find the co-ordinates of the point of contact.

**3.** Find the equation of the ellipse which touches the line $2x + 3y = 9$ and has the points $(-2, 0)$, $(2, 0)$ as its foci. Find also the co-ordinates of the point of contact of the line and the ellipse.

**4.** Show that the distance of the point $(a \cos \theta, b \sin \theta)$ from the focus $(ae, 0)$ of the ellipse $x^2/a^2 + y^2/b^2 = 1$ is $a(1 - e \cos \theta)$.

**5.** Prove that if the normals at $P$ $(6, 4)$ and $Q$ $(-8, 3)$ on the ellipse $x^2/100 + y^2/25 = 1$ meet at $G$, then the diameter through $G$ is perpendicular to $PQ$.

**6.** Find the equation of an ellipse of eccentricity $\frac{1}{2}$ which has a focus at $(3, 0)$ and $x = 1$ for corresponding directrix.

**7.** Find the equations of the tangents to the ellipse $x^2/9 + y^2/4 = 1$ which are parallel to the line $x = 2y$.

**8.** Find the equation of the ellipse of eccentricity $\frac{1}{2}$ which has its foci at the points $(-1, 0)$, $(1, 0)$. Find also the length of the latus rectum and verify that the tangent at either end of the latus rectum cuts the major axis on the directrix.

**9.** Find the equation of an ellipse which has the point $(2, 3)$ as an end of a latus rectum and its axes along the co-ordinate axes. At what point does the line $x - 2y = 8$ touch the ellipse?

**10.** Find the equation of an ellipse of eccentricity $0.8$ which has its centre at the origin and the lines $x = -25$ and $x = 25$ as directrices. Verify that the ellipse touches the line $9x + 20y = 300$.

**11.** The axes of an ellipse are the co-ordinate axes, its directrices pass through the points $(-5\frac{1}{3}, 0)$, $(5\frac{1}{3}, 0)$ and it touches the line $3x + 4y = 16$. Find its equation.

**12.** The major and minor axes of an ellipse lie along the lines $3x - 4y + 6 = 0$ and $4x + 3y - 17 = 0$ and the lengths of the semi-axes are 5 and 4. Find the eccentricity and the co-ordinates of the centre and foci.

**13.** Find the equation of an ellipse which has its axes along the co-ordinate axes and the line $3x - 2y = 5$ as the normal at the point $(15, 20)$.

**14.** Prove that, if the tangent at an end of the minor axis of an ellipse cuts the latus rectum produced in $D$, and $C$ is the centre, then a perpendicular to $CD$ through $D$ cuts the major axis on the directrix.

**15.** Prove that the tangent at the ends of the latera recta of the ellipse $x^2/a^2 + y^2/b^2 = 1$ form a quadrilateral of area $2a^2/e$, where $e$ is the eccentricity.

**16.** Prove that, if a series of ellipses have the same major axis, the tangents at the ends of their latera recta pass through one or other of two fixed points on the minor axis.

**17.** Find the equations of the tangents to the ellipse $4x^2 + 9y^2 = 180$ at the points $P$ (6, 2) and $P'$ (−6, −2). Find also the equations of the tangents that are parallel to the line $PP'$, and the co-ordinates of their points of contact.

**18.** Prove that the line $x + 3y = 9$ touches the ellipse $x^2/9 + y^2/8 = 1$, and find the co-ordinates of the point of contact. Find the co-ordinates of the foci of the ellipse, and verify that the product of the distances of the foci from the above tangent is equal to the square on the minor axis.

**19.** An ellipse has its foci at the points (−3, 0), (3, 0) and passes through the point $P$ (2, $2\sqrt{6}$). Prove that its eccentricity is $\frac{1}{2}$ and that the normal at $P$ passes through the point ($\frac{1}{2}$, 0).

**20.** An ellipse has a focus at the point (3, 0), the $y$-axis is the corresponding directrix and the point (6, 4) lies on the curve. Prove that the axes are in the ratio $6 : \sqrt{11}$.

**21.** Prove that, if the line $lx + my + n = 0$ is a normal to the ellipse $x^2/a^2 + y^2/b^2 = 1$, then $a^2/l^2 + b^2/m^2 = (a^2 - b^2)^2/n^2$.

**22.** Find the co-ordinates of the points on the ellipse $8x^2 + 25y^2 = 200$ at which the normals make angles of 60° with the major axis.

**23.** Find the values of $c$ for which the line $5x - 2y = c$ is normal to the ellipse $x^2 + 5y^2 = 9$.

**24.** Find the co-ordinates of the four points on the ellipse

$$9x^2 + 16y^2 = 1$$

the tangents at which are equally inclined to the co-ordinate axes; and prove that the normals at these points form a square of area 49/1800.

**25.** Find the equation of the normal at the point (2, 3) on the ellipse $3x^2 + 4y^2 = 48$ and the co-ordinates of the point in which the normal again cuts the curve. Show that the middle point of this normal chord is at a distance $\sqrt{73}/19$ from the centre of the ellipse.

**26.** Find the equation of an ellipse of eccentricity $\frac{1}{2}$ which touches the line $2x + 3y = 5$ and has its axes along the co-ordinate axes. Find the co-ordinates of the point of contact of the ellipse with the given line.

**27.** Find the equations of the two ellipses which have their axes along the co-ordinate axes, pass through the point $(2, 1)$ and touch the line $6x + 12y = 25$.

**28.** Prove that, if an ordinate $NP$ to an ellipse is produced to meet the tangent at the end of the latus rectum in $Q$, then $QN = SP$, where $S$ is the corresponding focus.

**29.** The tangent is drawn at the point $P$ $(2, 1)$ on the ellipse $4x^2 + 9y^2 = 25$ whose centre is $O$, and the diameter $DOD'$ of the ellipse is parallel to the tangent at $P$. Find the co-ordinates of $D$ and $D'$ and prove that the tangents at these points are parallel to the radius $OP$.

**30.** The tangent at $P$ to an ellipse meets a directrix in $T$ and $S$ is the corresponding focus. Prove that $PST$ is a right angle.

**31.** The foci of an ellipse are the points $(0, 0)$ and $(8, 6)$ and the eccentricity is $\frac{2}{5}$. Find the co-ordinates of the centre and the equations and lengths of the major and minor axes. Find also the equations of the directrices.

**32.** Show that, if the normal at a point $P$ $(x', y')$ on an ellipse of focus $S$ and eccentricity $e$ meets the major axis in $G$ and $GL$ is perpendicular to $SP$, then $GL = ey'$ and $PL =$ the semi-latus rectum.

**33.** $P$ denotes any point on an ellipse of which the major axis is $AA'$. Prove that, if $AP$, $A'P$ cut the minor axis in $M$, $M'$, then the tangent at $P$ bisects $MM'$.

### 9.6 The Hyperbola.

**1.** Prove that, if a variable line moves with its ends on the co-ordinate axes so as to enclose with them a constant area, then the locus of the middle point of the line is a rectangular hyperbola.

**2.** Prove that, if a straight line moves with its ends on the co-ordinate axes so as to form with them a triangle of constant area $c^2$, then the line touches the rectangular hyperbola $2xy = c^2$.

**3.** Find the value of $c^2$ if the rectangular hyperbola $xy = c^2$ touches the line $3x + 5y = 30$.

**4.** Show that the equation of the tangent at $(x', y')$ on the rectangular hyperbola $xy = c^2$ may be put in the form $x/x' + y/y' = 2$.

**5.** Prove that the line $3x + 4y = 24c$ touches the hyperbola $xy = 12c^2$. What is the area of the triangle whose sides are the asymptotes and this tangent?

**6.** Find the points in which the line $2x + y = 3c$ cuts the hyperbola $xy = c^2$, showing that one of the points of intersection lies on the transverse axis of the hyperbola.

**7.** Prove that, if $a$ and $b$ are real numbers of opposite signs, the straight line $ax + by = 1$ cannot touch the rectangular hyperbola $xy = c^2$.

**8.** Prove that the straight line $x + t^2y = 2ct$ is a tangent to the rectangular hyperbola $xy = c^2$, and that no perpendicular line can touch the curve.

**9.** Prove that, if through any point on the curve $xy = c^2$ two perpendicular lines are drawn, their other intersections with the curve will lie on opposite branches.

**10.** The normal at $P$ $(3, 4)$ on the rectangular hyperbola $xy = 12$ meets the curve again at $Q$. Prove that $PQ = 125/12$.

**11.** Find the equations of the tangent and normal at the point $x = 4ct$, $y = 3c/t$ on the rectangular hyperbola $xy = 12c^2$. Verify that the length of the tangent intercepted by the asymptotes is bisected at the point of contact.

**12.** Find the equation of the tangent at the point $x = 2ct$, $y = 5c/t$ on the hyperbola $xy = 10c^2$, and show that the area of the triangle the tangent forms with the asymptotes is independent of $t$.

**18.** Find the equation of the chord joining the points $(ct_1, c/t_1)$ and $(ct_2, c/t_2)$ on the rectangular hyperbola $xy = c^2$. Find the equation of the tangent parallel to the chord, and the co-ordinates of its point of contact.

**14.** Show that the hyperbolas $x^2 - y^2 = 20$ and $xy = 24$ cut at right angles at all their common points.

**15.** If the line $2x - ky = 8$ touches a rectangular hyperbola $xy = 9$, what is the value of $k$, and what are the co-ordinates of the point of contact?

**16.** Find tne co-ordinates of the foci and the equations of the directrices of the hyperbola $xy = c^2$, referred to the asymptotes as axes.

**17.** A circle cuts the rectangular hyperbola $xy = 1$ in the points $(x_r, y_r)$, $r = 1, 2, 3, 4$. Prove that $x_1 x_2 x_3 x_4 = y_1 y_2 y_3 y_4 = 1$.

**18.** Find the points of intersection of the line $8y - 2x = 15c$ and the rectangular hyperbola $xy = c^2$. Prove that the line is a normal to the curve at one of these intersections.

**19.** Prove that the normal at the point $P$ $(ct, c/t)$ on the rectangular hyperbola $xy = c^2$ meets the curve again at $Q$ so that

$$PQ = c \left(t^2 + 1/t^2\right)^{\frac{3}{2}}.$$

**20.** Four points are taken on a rectangular hyperbola $xy = c^2$. Find the condition that the chord joining two of the points should be perpendicular to the chord joining the other two. Prove that, if the condition is satisfied for one pair of such chords, then it is true for all three pairs.

**21.** Prove that the locus of the middle points of chords of the rectangular hyperbola $xy = c^2$ which pass through the point $(2c, 2c)$ is an equal hyperbola and find the co-ordinates of its vertices.

**22.** The radius $OP$ from the centre to a point $P$ on the rectangular hyperbola $xy = c^2$ makes an angle $\theta$ with the $x$-axis, and the normal at $P$ cuts the axes of $x$ and $y$ in $G$ and $G'$. Prove that $PG/PG' = \tan^2 \theta$.

# ANSWERS

**1.61**  1. $\sqrt{106}$.  2. (i) $\sqrt{13}$; (ii) $\sqrt{37}$; (iii) $\sqrt{\{2\,(a^2+b^2)\}}$.

7. $\frac{1}{2}\sqrt{34}$.  10. 8, $-1$.  11. $\frac{3}{2}\sqrt{2}$; 8; $\frac{3}{2}\sqrt{10}$.

12. $-25$, $-17\frac{1}{2}$; $-15$, $-15$; $-5$, $-12\frac{1}{2}$.

13. 5, 0;  15, 5.  15. 10; 15.  16. $-3$, $-4$; $-6$, 10.

17. 9.  18. 14, $-1$; $-16$, 9.

**1.9**  1. $3x^2 + 3y^2 - 4x - 14y + 15 = 0$.  2. $5x - 7y - 6 = 0$.

3. $x^2 + y^2 - 6x + 8y = 0$.  4. $y^2 - 4x - 2y + 5 = 0$.

5. $x^2 + y^2 - 7x - 7y = $ const.  6. $4x^2 + 3y^2 - 8y + 4 = 0$.

**2.123**  (i) $y = x + 1$.  (ii) $y = x\sqrt{3} - 3$.  (iii) $y = -x + 2$.

(iv) $y = \sqrt{3}\,(x - 2)$.  (v) $y + x + 2 = 0$.  (vi) $y + x + 1 = 0$.

**2.21**  1.  (i) 6, 4;  (ii) $-5$, $\frac{5}{3}$;  (iii) 0, 0;  (iv) $\dfrac{1}{a-b}$, $\dfrac{1}{a+b}$;

(v) $-\dfrac{c}{m}$, $c$.

2.  (i) $4x - 3y = 12$;  (ii) $6x - 5y = -80$;

(iii) $ax + by = 1$;  (iv) $x - y = 2a$.

3. $\dfrac{h}{x} + \dfrac{k}{y} = 1$.

**2.45**  1. $x\sqrt{3} + y = \pm\,6$.

2. $x\sqrt{3} - y = 2\,(1 + \sqrt{3})$;  $x \cos 330° + y \sin 330° = 1 + \sqrt{3}$.

3. $x + y = 5$;  $135°$.  4. $x + y = 12$.

5. $x = 2$;  $x - 2y + 6 = 0$;  $2x + 3y + 5 = 0$.

6. $5x - 3y + 2 = 0$;  $x + 12y - 8 = 0$;  $11x + 6y - 4 = 0$.

7. $4x - y - 11 = 0$.  8. $3 + \sqrt{3}$;  $-1 - \sqrt{3}$.

9. $7x + y - 17 = 0$;  $7x + y - 37 = 0$.

10. $7/\sqrt{17}$;  $4x - y = 14$.  11. $x + 2y + 5 = 0$.

12. $7x - 2y = 14$;  $7x - 2y = 23$.

13. $4x - 5y + 13 = 0$;  $4x - 5y = 0$.

14. $x + 3y - 2 = 0$;  $x + 3y - 11 = 0$.

15. $2x + 5y = -22$.

**2.53**  1. $-2$, 0;  $-42$, $-16$;  $-\frac{6}{11}$, $-\frac{24}{11}$.  2. $-68$; 43.

3. $22x - 11y = 53$.  4. $34x + 57y = 0$.

5. $60°$.  6. $90°$.  7. $7x - 5y = 34$;  $(\frac{52}{37}, -\frac{127}{37})$.

8. $bx - ay = 0$.  9. $2x + 5y = 0$.

10. $2y - x = 0$;  $2y - x = \pm\,3$.

11. $5x - y = 0$; $x - 5y = 0$; $x + y = 8$.

13. $x^2 + y^2 - 24x = 0$.

**2.66**    1. 1.      2. 2.      3. 0.      **4.** Opposite.

5. $6x - 6y = 1$;   $2x + 2y = 17$;   the first.

6. $2x + 3y = 7$.

7. $8x - 4y + 10 = \pm 5$;   $3x - 4y + 5 = 0$;   $3x - 4y + 15 = 0$; the first.

8. $\frac{22}{6}, 1\frac{3}{6}$;   $\frac{11}{2}, \frac{5}{2}$.        10.   $5x + 2y = 0$.

11. $5x + 3y + 8 = 0$.      12.   $x + y + 1 = 0$.

13. $\frac{1}{3}(x' + x'' + x''')$;   $\frac{1}{3}(y' + y'' + y''')$.

14. $k:l:m = 19:-11:17$;   $(2, 3)$.

15. $18x + 5y + 1 = 0$;   $6x + 11y - 5 = 0$.

16. $x - 3y - 3 = 0$;   $3x - 2y - 5 = 0$;   $2x + y - 2 = 0$.

18. $1\frac{13}{13}$;   $\frac{6}{13}$.

**2.72**    1.   (i) $x = 0$, $y = 0$;        (ii) $x + y = 0$, $x - y = 0$;

(iii) $3x - y = 0$, $2x + y = 0$;

(iv) $y - x(\tan\theta + \sec\theta) = 0$, $y - x(\tan\theta - \sec\theta) = 0$;

(v) $x - y = 0$, $x + (2 + \sqrt{8})y = 0$, $x + (2 - \sqrt{8})y = 0$.

2. (i) $60°$;   (ii) $90°$;   (iii) $\tan^{-1}\frac{2}{5}$;   (iv) $\tan^{-1}4$;   (v) $45°$.

**2.9**    1. $x - y + 1 = 0$.      2. $(-1, -1)$.      3. $\frac{23}{26}, \frac{41}{26}$.

4. $10x + 8y = 33$; $8x - 10y + 1 = 0$.

5. $x^2 + y^2 + x - 5y - 2 = 0$.      6. $(-3, 8)$.

8. $5, 8$.      9. $3\frac{1}{4}\frac{1}{0}\frac{1}{8}$.      10. $(\frac{3}{5}, \frac{31}{35})$.

11. $3y - x + 2a = 0$;   $3y - x - 3a = 0$;   $y + 3x - 6a = 0$;   $y + 3x - a = 0$.

15. $(ax + by + c)(a'x' + b'y' + c')$
$$- (a'x + b'y + c')(ax' + by' + c) = 0.$$

18. $-2, -1$.      19. $\theta$.

21. $60°$.      22. $\frac{7}{13}$.

23. $\sqrt{18}.(3x - 5y + 2) = \pm\sqrt{34}.(2x + 3y - 5)$.

24. $-10$;   $\tan^{-1}\frac{23}{2}$.      25. $-3$;   $-7$.

**3.12**    1. $x^2/a^2 + y^2/b^2 = 1$.      2. $x^2 + y^2 = $ const.

3. $x^2 + ky^2 = a^2$.      4. $x^2 - 2kxy - y^2 = a^2$.

**3.23**    1.   (i) $11x - y = 16$;      (ii) $y = 2$;

(iii) $y = 2x$;      (iv) $225x - 16y - 200 = 0$;

(v) $y = 0$;      (vi) $26x - y = 53$;

(vii) $3a^2x - y = 5a^3$;      (viii) $y - y' = 2(ax' + b)(x - x')$;

(ix) $y = 2$;      (x) $x - y + a = 0$.

2. $5x - y - 4 = 0$;   $3x + y + 4 = 0$;   $(0, -4)$.

**4.** $(2, \frac{14}{3})$; $(-2, -\frac{14}{3})$. **5.** $(3, 18)$; $(-3, 18)$.
**6.** $(1, \frac{8}{3})$; $(-1, -\frac{11}{3})$; $(-2, -\frac{40}{3})$.

**3.33** **1.** (i) $3x + 4y = 25$; (ii) $2x + 3y = 2$;

(iii) $y = x + a$; (iv) $\frac{x}{a}\cos\theta + \frac{y}{b}\sin\theta = 1$;

(v) $x\sec\theta - y\tan\theta = a$; (vi) $x + y - 4 = 0$;
(vii) $4x + 3y - 15 = 0$; (viii) $y = a$.

**3.43** **1.** (i) $x - y + 1 = 0$; (ii) $4x - 5y + 6 = 0$;
(iii) $x + y = 3a$; (iv) $ax\sec\theta - by\csc\theta = a^2 - b^2$;
(v) $2x\sin\theta + y = 5a\tan\theta$; (vi) $x = a$.
**11.** $5x - 4y - 6a = 0$.

**3.51** (i) $x^2 + y^2 = a^2$. (ii) $y^2 = 4ax$.
(iii) $\frac{x^2}{a^2} + \frac{y^2}{b^2} = 1$. (iv) $\frac{x^2}{a^2} - \frac{y^2}{b^2} = 1$.

(v) $xy = c^2$. (vi) $\left(\frac{x}{a}\right)^{\frac{2}{3}} + \left(\frac{y}{b}\right)^{\frac{2}{3}} = 1$.

(vii) $(x^2 + y^2 - b^2)^2 = a^2\{(x+b)^2 + y^2\}$.

**3.53** **2.** (i) $x\cos t + y\sin t = a$; $y = x\tan t$.
(ii) $x - yt + at^2 = 0$; $tx + y = at^3 + 2at$.

(iii) $\frac{x}{a}\cos t + \frac{y}{b}\sin t = 1$; $ax\sec t - by\csc t = a^2 - b^2$.

(iv) $\frac{x}{a}\sec t - \frac{y}{b}\tan t = 1$; $ax\cos t + by\cot t = a^2 + b^2$.

(v) $x + yt^2 - 2ct = 0$; $t^3x - ty - c(t^4 - 1) = 0$.

(vi) $\frac{x}{a}\sin t + \frac{y}{b}\cos t = \sin t\cos t$;

$ax\cos t - by\sin t = a^2\cos^4 t - b^2\sin^4 t$.

(vii) $x(a\cos t + 2b\cos 2t) + y(a\sin t + 2b\sin 2t)$
$= a^2 + 2b^2 + 8ab\cos t$;
$x(a\sin t + 2b\sin 2t) - y(a\cos t + 2b\cos 2t) = ab\sin t$.

**3.63** **1.** $x^2 - y^2 = 1$. **2.** $(-1, 2)$.
**4.** $x^2 - y^2 = a^2$. **5.** $y^2 = 4ax$.
**6.** $(-1, -2)$; $\frac{1}{2}\tan^{-1}\frac{4}{3}$. **7.** $22\frac{1}{2}°$.

**4.16** **1.** (i) $\frac{2}{\sqrt 3}$; $2, -1$; (ii) $\sqrt{(a^2 + b^2)}$; $0, 0$;

(iii) $\frac{1}{\sqrt 2}$, $-4, 1$; (iv) $2$; $\frac{1}{2}, \frac{1}{2}$.
**2.** $3x^2 + 3y^2 - 5x - 5y = 0$.
**3.** $x^2 + y^2 - 13x - 13y + 52 = 0$.

    **4.** $x^2 + y^2 - 2ax - 2by = 0$.

    **5.** $2, 1$;  $x^2 + y^2 - 4x - 2y - 5 = 0$.

    **6.** $(x - a)^2 + (y - b)^2 = (p \pm c - x \cos \alpha - y \sin \alpha)^2$.

**4.42**  **1.** Outside;  inside;  inside.

    **2.** $(2, 3)$;  $(\frac{4\,8}{1\,3}, \frac{9}{1\,3})$;  $\pm 13$.    **3.** $8, 8$.        **4.** $1, -\frac{1}{2}$.

    **6.** $x^2 + y^2 - 2x - 2y + 1 = 0$;  $x^2 + y^2 - 10x - 10y + 25 = 0$.

    **7.** $5$.                **8.** $2(x + y) = 7 \pm \sqrt{10}$.

    **9.** $x + y - 7 = 0$;  $x + y - 5 = 0$.

    **10.** $3x + 4y - 26 = 0$;  $4x - 3y - 18 = 0$.

    **11.** $x + y = 13$;  $x - y = 7$;  $x - y = -5$.

    **12.** $gx + fy = 0$.

    **14.** $(3, 4)$;  $(4, -3)$;  $3x + 4y = 25$;  $4x - 3y = 25$.

    **15.** $3, -1$.          **16.** $x^2 + y^2 \pm 2\sqrt{5}x - 4y + 5 = 0$.

    **17.** $\frac{1}{2}$;  $\frac{7}{2}$.         **18.** $x - y - 3 = 0$.

    **20.** $(al + bm + n)^2 = (l^2 + m^2)\, r^2$.

**4.57**  **1.** (i) $2x - y = 6$;  (ii) $8x + 4y = 6$;  $2\frac{8}{11}$, $-\frac{6}{11}$.

    **2.** (i) $(3\frac{4}{7}, 5\frac{1}{2})$;   (ii) $(-7\frac{1}{2}, 1\frac{1}{2})$;  $17x - 53y + 207 = 0$.

    **5.** $6x + 5y - 48 = 0$.        **6.**  (i) $\left(-\dfrac{a^2 l}{n}, -\dfrac{a^2 m}{n}\right)$;

        (ii) $\left(\dfrac{f(lf - mg) + gn - cl}{gl + fm - n}, \dfrac{g(mg - lf) + fn - cm}{gl + fm - n}\right)$.

**4.8**  **1.** $x^2 + y^2 - 20x + 20y + 55 = 0$.

    **2.** $x^2 + y^2 + 6x + 4y - 4 = 0$.

    **3.** $31x^2 + 31y^2 + 90x - 12y - 109 = 0$.

    **4.** (i) $x - y - 2 = 0$;  (ii) $6x - 7y - 1 = 0$;  (iii) $gx - fy = 0$.

    **5.** $x^2 + y^2 - 5x + 6y - 1 = 0$.

    **6.** $-\frac{1}{8}$;  $\frac{5}{4}$.          **7.** $(1, -1)$;  $(5, -1)$.

    **10.** $x^2 + y^2 + 2x - 4 = 0$;  $15x^2 + 15y^2 + 10x + 20y - 48 = 0$.

    **11.** $(1\frac{3}{22}, \frac{23}{22})$;  $\dfrac{\sqrt{2766}}{22}$.      **12.** $x^2 + y^2 - 2x - 7y - 9 = 0$.

**4.9**  **1.** A circle.         **2.** A circle.        **3.** A circle.

    **4.** $ab = cd$;  $x^2 + y^2 - (a + b)x - (c + d)y + ab = 0$.

    **5.** $x = 0$, $y = 0$;  $(b^2 - a^2)x + 2aby = 2ab^2$;

    $2abx + (a^2 - b^2)y = 2a^2 b$.

    **6.** $x \cos \theta + y \sin \theta = a$.

    **7.** $(x^2 + y^2 + c^2)(1 - k^2) - 2cx(1 + k^2) = 0$.      **8.** $45°$.

    **11.** $(lmf - m^2 g + l)/(l^2 + m^2)$;  $(lmg - l^2 f + m)/(l^2 + m^2)$.

    **21.** $x \cos \alpha + y \sin \alpha = a$;  $\{a(1 \pm \sqrt{2})\cos \alpha, a(1 \pm \sqrt{2})\sin \alpha\}$.

**5.61**  **1.** $(\frac{1}{4}a, a)$.         **2.** $(4a, 4a)$.

**8.** $(\frac{1}{16}a, \ -\frac{1}{2}a)$; $(\frac{1}{4}a, \ a)$; $(-\frac{1}{8}a, \ \frac{1}{4}a)$.   **4.** $2\sqrt{3}y = 3x + 4a$.

**10.** $\left\{a\left(t + \dfrac{2}{t}\right)^2, \ -2a\left(t + \dfrac{2}{t}\right)\right\}$.

**5.26**   **1.** (i) $2\frac{2}{3}$, $\dfrac{\sqrt{5}}{3}$, $(\pm\sqrt{5}, \ 0)$;   (ii) $2\frac{2}{3}$, $\dfrac{\sqrt{5}}{3}$, $(0, \ \pm\sqrt{5})$;

(iii) $2\sqrt{2}$, $\dfrac{1}{\sqrt{2}}$, $(\pm 2, \ 0)$.

**2.** $3x^2 + 4y^2 = 108$; $x = \pm 12$.

**3.** $7(x-4)^2 + 16y^2 = 112$; $3\frac{1}{2}$.

**4.** $28(x-6)^2 + 64y^2 = 567$.

**5.** $3(x-2)^2 + 4(y-1)^2 = 12$; $4$; $2\sqrt{3}$.

**6.** $7x^2 - 2xy + 7y^2 - 22x - 38y + 31 = 0$.

**7.** Inside; outside.     **8.** $3\frac{1}{3}, \ -1\frac{1}{3}$.     **9.** $\pm\sqrt{2}$.

**10.** $x + 2y = 4$; $y = 2x \pm \sqrt{34}$.     **11.** $y = \sqrt{3}x \pm \sqrt{\frac{5}{2}}$.

**12.** $(ae, \ a(1 - e^2))$.

**6.34**   **1.** $10x + 21y = 6$.     **2.** $(1, \ 1)$.

**3.** $(12, \ -1\frac{1}{4})$; $(1\frac{1}{2}, \ 1\frac{1}{8})$.     **5.** $9x - 14y + 1 = 0$.

**6.56**   **2.** $a\cos\alpha, a\sin\alpha$; $\dfrac{x}{a}\cos\alpha + \dfrac{y}{b}\sin\alpha = 1$; $x\cos\alpha + y\sin\alpha = a$.

**8.9**   **1.** (i) $r^2 + 2r(g\cos\theta + f\sin\theta) + c = 0$;

(ii) $r = a\cos\theta$;          (iii) $r^2\cos 2\theta = c^2$;

(iv) $r^2\sin 2\theta = c^2$;          (v) $r^3\cos 3\theta = a^3$.

**2.** (i) $y + a = 0$;          (ii) $x^2 + y^2 = ay + bx$;

(iii) $x^2 - y^2 = a^2$;          (iv) $x^2 + y^2 = (l - ex)^2$;

(v) $3x^2y - y^3 = a^3$;

(vi) $(x^2 + y^2)^2 = a^2(x^2 + y^2) + b^2(x^2 - y^2)$.

**3.** (i) $pr = 2c^2$;          (ii) $pa^2 = r^3$;

(iii) $pa^n = r^{n+1}$;          (iv) $\dfrac{1}{p^2} = \dfrac{1}{r^2} + \dfrac{1}{a^2}$.

**9.1**   **1.** $10$; $(9, \ 11\frac{1}{2})$.

**2.** $x - 3y + 9 = 0$, $6x + 5y + 8 = 0$, $8x - y = 20$.

**3.** $5x - 6y + 9 = 0$, $x + 8y = 13$, $3x + y = 2$.

**4.** $13, \ 5x - 12y + 26 = 0$, $2\frac{1}{13}$, $10\cdot5$.     **5.** $3, \ 1\frac{1}{6}, \ 1\frac{4}{5}$.

**6.** $5x + 4y + 25 = 0$.     **7.** $2x + 7y - 9 = 0$, $7x - 2y - 5 = 0$.

**8.** $16$.     **10.** $2/\sqrt{13}, \ 1/\sqrt{13}$.     **11.** $(5, \ 4\frac{1}{2})$, $15$.

**13.** $9x - y = 25$, $x + 9y = 21$.

**14.** $18x - 4y + 51 = 0$.     **15.** $(12\frac{14}{25}, \ -9\frac{2}{25})$.

**16.** $77x - 99y + 138 = 0$, $14x + 112y - 177 = 0$, $7x + y - 3 = 0$,
$(1\frac{59}{770}, \ \frac{17}{110})$.

**17.** $(3\frac{2}{5}, \ 2\frac{1}{5})$, $3x - y = 8$, $x + 3y = 8$.     **18.** $(\frac{47}{60}, \ \frac{269}{60})$.

**19.** $39x + 65y = 107\cdot 65x - 39y + 71 = 0$.

20. $27x + 13y - 14 = 0.$      21. $(\frac{30}{13}, \frac{59}{13}).$

22. $(0, 5).$     23. $\pm 3\frac{2}{3}.$

24. $x = y (\operatorname{cosec} \theta \pm \cot \theta),\ \frac{1}{2}\pi \pm \theta.$

**9.2**
1. $y = 2x - 2,\ x + y = 1,\ \tan^{-1} 3.$     2. $(2, \frac{2}{3}).$

4. $(2, 10),\ (-2, -10).$     5. $10x - y = 13,\ (\frac{19}{60}, -\frac{533}{400}).$

6. $x + 5y + 13 = 0,\ 15.$     7. $x - y + 3 = 0,\ (5, -28).$

9. $4x - y = 7,\ x + 4y = 6;\ 2\frac{1}{8}.$

10. $1,\ x = \pm \sqrt{(2/3)}.$     11. $0,\ 1;\quad (1, 0),\ (\frac{5}{3}, -\frac{4}{27}).$

12. $(1, 1),\ (-\frac{1}{3}, \frac{23}{27});\quad x - y = 0,\ x + y = 2,\ 27\,(y - x) = 32,$
    $27\,(x + y) = 14.$

13. $(1, 0),\ (-\frac{1}{3}, 0),\ (0, 1);\quad -4, 4, 2;\quad (\frac{1}{3}, \frac{4}{3}).$

14. $4x\,(x^2 - 1);\quad -1, 0, 1;\quad (2, 9),\ (-2, 9).$

15. $-16, 0, 16;\quad (-\sqrt{2}, -4),\ (0, 0),\ (\sqrt{2}, -4).$

16. $2x - 3;\quad (\frac{3}{2}, -\frac{25}{4});\quad 5, -5;\quad x = \frac{3}{2};\quad 4y = 4x^2 - 25.$

17. $2;\quad (3, -2),\ (-3, 2).$

18. $11x^2 + 96xy + 39y^2 + 52x - 14y - 109 = 0.$

19. $19x^2 - 4xy + 16y^2 + 6x + 12y - 9 = 0.$

20. $x^2 + y^2 - 2x - 3y = 0.$

21. $x^2 + y^2 + 8x = 0.$     22. $\dfrac{1}{x} + \dfrac{1}{y} = 2.$

23. $x^2 - y^2 = 4;\quad (t^2 + 1)\,x - (t^2 - 1)\,y = 4t.$

**9.3**
1. $3x + 4y = 0,\ 3x + 4y + 50;\quad (6, 0),\ (0, 8).$

2. $x^2 + y^2 - 3x + 2y - 1 = 0,\ (-\frac{1}{2}, -\frac{1}{2}),\ (\frac{7}{2}, -\frac{5}{2}).$

3. $4x - 3y = 5;\quad (\frac{5}{4}, 0),\ (0, \frac{5}{3});\quad 4x - 3y = 5,\ -4x + 3y = 5.$

4. $0, -\frac{4}{3};\quad (-1, 0),\ (\frac{3}{5}, -\frac{4}{5}).$

5. $3x + 4y - 47 = 0,\ 3x + 4y + 3 = 0.$

6. $9x - 13y + 142 = 0.$

7. $x^2 + y^2 - 11x - 10y + 24 = 0,\ (8, 0).$

8. $(2, 3),\ (3, 2);\quad x + y = 2,\ x + y = \frac{16}{3}.$

9. $(5, 1),\ (3, 2),\ 3x + 11y = 26,\ 7x + 9y = 39,\ (\frac{39}{10}, \frac{13}{10}).$

11. $(\frac{21}{25}, -\frac{28}{25}),\ (3, -4),\ (-3, 4).$     12. $(\frac{24}{13}, -\frac{84}{13}).$

13. $x - 2y + 4 = 0,\ x - 2y = 6.$     14. $\tan^{-1} 2 \cdot 4.$

15. $x^2 + y^2 + 10x - 10y + 25 = 0,\ x^2 + y^2 + 2x - 2y + 1 = 0.$

16. $x^2 + y^2 - 4x - 6y - 5 = 0.$

17. $2x + 3y = 14,\ 2x + 3y + 12 = 0.$     18. $x + 2y = 5.$

19. $20x^2 + 20y^2 - 60x - 60y + 81 = 0.$

20. $x^2 + y^2 + 6x - 10y - 51 = 0.$     21. $x^2 + y^2 = 2c.$

**9.4**
1. $\frac{1}{2}a.$     2. $y^2 = 8x;\quad (2, 0),\ (18, \pm 12).$

3. $(\frac{3}{8}, -\frac{3}{2}).$     4. $12,\ (\frac{4}{3}, 8).$

5. $(at^2, 2at),\ (0, at).$       8. $y^2 = 4x;\quad (\frac{5}{2}, 8).$

13. $(a/t^2, -2a/t)$.      **14.** $x = a\left(t + \dfrac{2}{t}\right)^2, y = -2a\left(t + \dfrac{2}{t}\right).$

15. $2$;   $2x + y = 24$, $(18, -12)$.

16. $x - 3y + 36 = 0$, $9x + 3y + 4 = 0$, $x = -4$.

17. $(36, 12), (0, 6)$.    **18.** $24x - 7y + 3 = 0$.    **19.** $1$.

**9.5**

1. $x^2/16 + y^2/12 = 1$.          2. $(2, 3)$.

3. $x^2/9 + y^2/5 = 1$.   $(2, \tfrac{5}{3})$.      6. $3x^2 + 4y^2 - 22x + 35 = 0$.

7. $x + 2y = \pm 5$.          8. $x^2/4 + y^2/3 = 1$;   3.

9. $3x^2 + 4y^2 = 48$;   $(2, -3)$.    10. $9x^2 + 25y^2 = 3600$.

11. $7x^2 + 16y^2 = 112$.

12. $\tfrac{3}{5}$, $(2, 3)$, $(4\cdot 4, 4\cdot 8)$, $(-\cdot 4, 1\cdot 2)$.      13. $x^2/9 + y^2/8 = 75$.

17. $4x + 3y = \pm 30$;   $x - 3y = \pm 15$;   $(3, -4), (-3, 4)$.

18. $(1, 2\tfrac{2}{3})$;   $(1, 0), (-1, 0)$.     22. $(\pm 25/7, \pm 8\sqrt{3}/7)$.

23. $\pm 8$.               24. $(\pm \tfrac{4}{15}, \pm \tfrac{3}{20})$.

25. $2x - y = 1$;   $(-\tfrac{22}{19}, -\tfrac{63}{19})$.

26. $24x^2 + 27y^2 = 50$;   $(\tfrac{5}{6}, \tfrac{10}{9})$.

27. $4x^2 + 9y^2 = 25$, $9x^2 + 64y^2 = 100$.      29. $(\pm \tfrac{3}{2}, \mp \tfrac{4}{3})$.

31. $(4, 3)$; $3x - 4y = 0, 4x + 3y = 25$; $12\cdot 5, 7\cdot 5$; $64x + 48y = 1025$, $64x + 48y + 225 = 0$.

**9.6**

3. $15$.            5. $24c^2$.            6. $(c, c)$, $(\tfrac{1}{2}c, 2c)$.

11. $3x + 4t^2y - 24ct = 0$, $4t^3x - 3ty + 9c - 16ct^4 = 0$.

12. $5x + 2t^2y - 20ct = 0$.

13. $x + t_1 t_2 y = c(t_1 + t_2)$, $x + t_1 t_2 y = 2c\sqrt{(t_1 t_2)}$;   $(c\sqrt{(t_1 t_2)}, c/\sqrt{(t_1 t_2)})$.      15. $-\tfrac{1}{8}$;   $(\tfrac{3}{4}, 12)$.

16. $(\sqrt{2}c, \sqrt{2}c)$, $(-\sqrt{2}c, -\sqrt{2}c)$, $x + y = \pm \sqrt{2}c$.

18. $(c/2, 2c)$, $(-8c, -c/8)$.      20. $t_1 t_2 t_3 t_4 = -1$.

21. $(x - c)(y - c) = c^2$, $(2c, 2c)$, $(0, 0)$.